Studies in Computational Intelligence

Volume 1220

The series "Studies in Computational Intelligence" (SCI) publishes new developments and advances in the various areas of computational intelligence—quickly and with a high quality. The intent is to cover the theory, applications, and design methods of computational intelligence, as embedded in the fields of engineering, computer science, physics and life sciences, as well as the methodologies behind them. The series contains monographs, lecture notes and edited volumes in computational intelligence spanning the areas of neural networks, connectionist systems, genetic algorithms, evolutionary computation, artificial intelligence, cellular automata, self-organizing systems, soft computing, fuzzy systems, and hybrid intelligent systems. Of particular value to both the contributors and the readership are the short publication timeframe and the world-wide distribution, which enable both wide and rapid dissemination of research output.

Indexed by SCOPUS, DBLP, WTI AG (Switzerland), zbMATH, SCImago.

All books published in the series are submitted for consideration in Web of Science.

Oluwatosin Ahmed Amodu ·
Raja Azlina Raja Mahmood · Huda Althumali ·
Umar Ali Bukar · Nor Fadzilah Abdullah ·
Chedia Jarray

Markov Decision Processes and Reinforcement Learning for Timely UAV-IoT Data Collection Applications

Springer

Oluwatosin Ahmed Amodu
Wireless Lab
Department of Electrical and Electronics Engineering
Faculty of Engineering and Built Environment
Universiti Kebangsaan Malaysia
Bangi, Selangor, Malaysia

Huda Althumali
Department of Computer Science
College of Science and Humanities
Imam Abdulrahman Bin Faisal University
Jubail, Saudi Arabia

Nor Fadzilah Abdullah
Department of Electrical, Electronic and Systems Engineering
Faculty of Engineering and Built Environment
Universiti Kebangsaan Malaysia
Bangi, Selangor, Malaysia

Raja Azlina Raja Mahmood
Department of Communication Technology and Network
Faculty of Computer Science and Information Technology
Universiti Putra Malaysia
Serdang, Selangor, Malaysia

Umar Ali Bukar
Faculty of Information Science and Technology
Multimedia University
Beruang, Malaysia

Chedia Jarray
Bimatech
Medenine, Tunisia

ISSN 1860-949X ISSN 1860-9503 (electronic)
Studies in Computational Intelligence
ISBN 978-3-031-97010-8 ISBN 978-3-031-97011-5 (eBook)
https://doi.org/10.1007/978-3-031-97011-5

This Springer imprint is published by the registered company Springer Nature Switzerland AG
The registered company address is: Gewerbestrasse 11, 6330 Cham, Switzerland

Preface

The convergence of Unmanned Aerial Vehicles (UAVs) and the Internet of Things (IoT) has brought a new era of data-driven decision-making. This synergy presents unprecedented opportunities for addressing complex challenges across various domains, from precision agriculture to disaster response. At the heart of these applications lies the efficient collection and processing of fresh data.

This book aims to provide examples on the application of Markov Decision Processes (MDPs) to modeling and optimising UAV-assisted IoT data collection systems, as well as various adopted Deep Reinforcement Learning (DRL) algorithms in finding the optimal solutions to some of these MDP-based problems. The content is structured to provide a gradual progression, starting with an introduction to the UAV-IoT systems and their critical applications, followed by a discussion on the most important metric in determining data freshness, namely Age of Information (AoI) in Chap. 1. With the primary goal to minimise the AoI, the fundamentals of MDPs and their applications to model UAV-assisted WSN/IoT systems with varying optimisation objectives are then covered in Chap. 2. Subsequently, Chap. 3 investigates the DRL algorithms adopted to solve some of the MDP-based problems discussed in the previous Chapter. Finally, some of the key challenges in implementing DRL solutions for AoI-aware UAV-IoT applications are addressed in Chap. 4. The Chapters provided in this book will enable readers to grasp the core concepts of MDPs via the provided examples serving as a valuable resource for academic and industry professionals interested in exploring the frontiers of UAV-assisted IoT applications.

Bangi, Malaysia — Oluwatosin Ahmed Amodu
Serdang, Malaysia — Raja Azlina Raja Mahmood
Jubail, Saudi Arabia — Huda Althumali
Beruang, Malaysia — Umar Ali Bukar
Bangi, Malaysia — Nor Fadzilah Abdullah
Medenine, Tunisia — Chedia Jarray
April 2025

Acknowledgements

This book is intended to support students, researchers, and industry practitioners exploring the application of Markov Decision Processes and Reinforcement Learning algorithms in controlling Unmanned Aerial Vehicles for timely data collection in Wireless Sensor Networks (WSNs) and Internet of Things (IoT) applications. We hope the insights and findings presented here will inspire further innovation in time-sensitive, UAV-assisted IoT systems.

We also acknowledge the contributions of numerous researchers and practitioners whose work on MDPs and Reinforcement Learning algorithms has laid the foundation for the concepts described in this book. We hope it will serve as a motivation for further advancements in this exciting field.

The authors gratefully acknowledge the support of their respective institutions and organisations. Particularly, Oluwatosin Ahmed Amodu gratefully acknowledges the support of Universiti Kebangsaan Malaysia (UKM), particularly the Wireless Research@UKM, Faculty of Engineering and Built Environment, and the Center for Research and Instrumentation Management (CRIM), for the postdoctoral research fellowship and resources that facilitated the initial compilation of this book.

Contents

1 Introduction to AoI in UAV-Assisted Sensor and IoT Systems 1
1.1 Introduction 1
1.1.1 Age of Information 2
1.1.2 Deep Reinforcement Learning Applications 3
1.2 UAV-Assisted WSN/IoT Architecture and Assumptions 4
1.3 UAV-Assisted Sensor and IoT Applications 6
1.3.1 Industrial and Structural Monitoring 6
1.3.2 Environmental Monitoring 7
1.3.3 Precision Agriculture 8
1.3.4 Smart Cities 10
1.3.5 Disaster Management 11
1.4 Age of Information (AoI) 12
1.4.1 Factors Affecting Timely Data Reception in UAV-IoT Systems 14
1.4.2 Important Considerations in AoI-Aware UAV-IoT Design and Optimisation 16
1.5 Summary 20
References 21

2 AoI-Aware UAV-IoT Modeling Using MDPs 27
2.1 Introduction to Markov Decision Processes (MDPs) 27
2.2 Fully Observable MDP-Based Modeling and Problem Formulation 29
2.2.1 Trajectory Optimisation 32
2.2.2 Trajectory and Energy Optimisation 37
2.2.3 Trajectory and Scheduling Optimisation 46
2.2.4 Trajectory, Energy, and Scheduling Optimisation 53
2.2.5 Trajectory, Scheduling, and RIS Phase Shift Optimisation 57

2.2.6 Scheduling Optimisation and Altitude Control 59
2.2.7 Scheduling Optimisation and Packet Drop Rate 66
2.3 Partially Observable MDP-Based Modeling and Problem Formulation 68
2.3.1 Example 1 70
2.3.2 Example 2 71
2.3.3 Example 3 73
2.4 Summary 75
References 76

3 Reinforcement Learning and Deep Reinforcement Learning for AoI-Aware UAV-IoT 79
3.1 Reinforcement Learning and Deep Reinforcement Learning 79
3.1.1 Bellman Equation 80
3.1.2 Classes of RL Algorithms 82
3.1.3 Multi-Agent Deep Reinforcement Learning 88
3.2 DRL-Based Solutions for AoI-Aware UAV-IoT Applications 90
3.2.1 Examples of Value-Based Solutions and Algorithms 90
3.2.2 Examples of Policy-Based Solutions and Algorithms 108
3.2.3 Examples of Actor-Critic-Based Solutions and Algorithms 116
3.3 Summary 126
References 126

4 Challenges and Future Considerations 129
4.1 DRL-Related Challenges in Age of Information (AoI)-Aware UAV-IoT 129
4.1.1 Convergence 129
4.1.2 Complexity 130
4.1.3 Computation Overhead 131
4.1.4 Training Overhead 131
4.1.5 State-Action Space Dimension 132
4.1.6 Exploration 132
4.1.7 Stability 133
4.1.8 Choice of Parameters and Hyperparameters 133
4.2 AoI-Related Challenges in UAV-IoT Architecture 134
4.2.1 Reliability of Sensor Nodes' Connections 134
4.2.2 Optimal Scheduling Policies 134
4.2.3 Traffic Generation Pattern Adaptation 135
4.2.4 Choice of Algorithms and Trade-Offs 136
4.2.5 Nature of the Optimisation Problem 136
4.2.6 Optimisation Algorithms 137
4.2.7 Environmental Challenges 137
4.2.8 Proper Choice of Channel Models 137
4.2.9 Security Threats 138

4.2.10 Fine-Grained UAV Assumptions 138
4.2.11 Network Architecture and Practical Constraints 139
4.2.12 Real Deployments 139
4.3 Summary 139
References 140

Acronyms

A2C	Advantage Actor Critic
A2G	Air-to-Ground
A3C	Asynchronous Advantage Actor Critic
AC	Ant Colony
AoI	Age of Information
AP	Affinity Propagation
ASoA	Average-Sum-AoI
BS	Base Station
CH	Cluster Head
CP	Collection Point
CSI	Channel State Information
DC	Data Center
DDPG	Deep Deterministic Policy Gradient
DDQN	Double DQN
D3QN	Dueling Double Deep Q-Network
DNN	Deep Neural Network
DP	Dynamic Programming
DPG	Deterministic Policy Gradient
DQN	Deep Q Network
DRL	Deep Reinforcement Learning
EE	Energy Efficiency
EH	Energy Harvesting
ESA	Expected Sum AoI
FANETs	Flying Ad Hoc Networks
FDRL	Federated Deep Reinforcement Learning
FIFO	First-In-First-Out
GA	Genetic Algorithm
GNSS	Global Navigation Satellite System
GPU	Graphics Processing Units
GSDRL	Guided Search DRL
GTSP	Generalized Travelling Salesman Problem

IIoT	Industrial Internet of Things
IoT	Internet of Things
IoTD	Internet of Things Devices
LIFO	Last-In-First-Out
LoRA	Long Range
LOS	Line-of-Sight
LPWAN	Low Power Wide Area Network
MAC	Medium Access Control
MADRL	Multi-Agent DRL
MARL	Multi-Agent Reinforcement Learning
MDP	Markov Decision Process
NWAoI	Normalized Weighted Sum of Age of Information
OFDMA	Orthogonal Frequency-Division Multiple Access
POMDP	Partially Observable Markov Decision Process
PPO	Proximal Policy Optimisation
QoS	Quality of Service
RF	Radio Frequency
RL	Reinforcement Learning
SAC	Soft Actor Critic
SAGIN	Space-Air-Ground Integrated Network
SARSA	State–Action–Reward–State–Action
SGA	Stochastic Gradient Ascent
SMDP	Semi-Markov Decision Process
SN	Sensor Node
SNR	Signal-to-Noise Ratio
TD3	Twin-Delayed Deterministic Policy Gradient
TDMA	Time-Division Multiple Access
TRPO	Trust Region Policy Optimisation
UAV	Unmanned Aerial Vehicle
VDN	Value Decomposition Networks
VoI	Value of Information
VPO	Variational Policy Optimisation
WIT	Wireless Information Transfer
WPT	Wireless Power Transfer
WSN	Wireless Sensor Networks

Chapter 1
Introduction to AoI in UAV-Assisted Sensor and IoT Systems

1.1 Introduction

Unmanned Aerial Vehicle (UAV), or drone, is an aerial platform that operates without the need for an onboard human pilot. It can be remotely controlled by a ground-based operator or navigate autonomously or semi-autonomously based on a predefined flight plan. Today, UAVs are used in military operations, sensor data collection, precision agriculture, disaster response, environmental monitoring, smart cities, security and safety enforcement, target tracking, geographical mapping and surveying, fault detection, and object recognition, among others [1].

Due to their mobility, agility, and versatility, UAVs offer high-quality and cost-effective wireless communication solutions for a broad range of real-world applications. They can function as aerial base stations, providing on-demand wireless connectivity to targeted locations, or serve as aerial user equipment, coexisting with terrestrial users to enable services such as surveillance and delivery. UAVs are particularly valued for their ability to dynamically adjust altitudes, avoid obstacles, and improve line-of-sight communication, making them ideal for deployment in hotspots, remote regions, and hard-to-access environments [2]. Furthermore, UAVs are increasingly being utilized to support fast Internet access, local content caching, and edge computing services, driving forward the future of flexible and intelligent wireless networks [3].

UAVs are often deployed to support terrestrial Internet of Things (IoT) systems that involve sensors. In broad terms, a sensor is often described as a device that receives and responds to a stimulus or signal [4]. In more specific terms, a sensor node is a device, module, or subsystem that can detect events or changes in certain environments and send the information to other technological devices, such as a computer processor. Sensors interact with other technologies as a component that depends on other entities or as an embodiment of other technologies to solve specific problems. Examples of sensors commonly used in IoT systems include pressure sensors, temperature sensors, and proximity sensors, all of which play a critical role

O. A. Amodu et al., *Markov Decision Processes and Reinforcement Learning for Timely UAV-IoT Data Collection Applications*, Studies in Computational Intelligence 1220,
https://doi.org/10.1007/978-3-031-97011-5_1

in enabling real-time data acquisition and intelligent system responses [5]. A sensor network is a network composed of many smart devices (nodes) that are spatially distributed to perform an application-oriented task [6]. The IoT paradigm facilitates communication between electronic devices and sensors through the Internet to achieve its desired objective [7]. An IoT architecture can involve a physical, virtual, or hybrid system involving several active physical *things*, sensors, actuators, cloud services, protocols, users, communication layers, among others [8].

UAV-assisted IoT applications are abundant in our daily lives. For instance, monitoring and controlling pests on the farm, monitoring, sensing, and data collection in rural and urban areas, object tracking, monitoring air quality in our surroundings, disaster management and emergency response, video surveillance for security, and large-scale crowdsensing initiatives. They can also be integrated to support vehicular applications, intelligent transport systems, marine applications, and several time-constrained applications. The utilisation of unmanned aerial vehicles for data collection in large-scale wireless sensor networks (WSNs) and IoT applications has been of immense interest in practice due to its potential to conserve the energy of sensor nodes. Also, UAV's rapid mobility and maneuverability make them ideal for rural and urban monitoring, where they deliver sensed data from remote areas to base stations. This capability is vital in time-critical scenarios, such as medical supply delivery and disaster response, where timely data delivery can be a matter of dire necessity.

1.1.1 Age of Information

The Age of Information (AoI) has emerged as a key performance metric to quantify the timeliness of information updates in wireless communication systems, particularly those involving data transfer from a source to a destination, such as base stations or data centers in UAV-assisted IoT applications. Unlike traditional latency metrics that only measure the delay of individual packets, AoI captures the freshness of the received information, making it highly relevant for real-time monitoring, control, and decision-making applications. Ensuring timely data collection and delivery requires extensive planning and network optimisation, especially in dynamic environments.

However, achieving optimal AoI performance becomes increasingly challenging as the network scale grows and the complexity of operational alternatives increases, making traditional optimisation algorithms often impractical due to their high computational overhead and inability to adapt in real time. In the context of UAV-assisted sensor and IoT applications, AoI represents the age of the latest information collected by UAVs from distributed sensor nodes or IoT devices. Maintaining a low AoI in these networks is crucial to ensure that the collected data accurately reflects the current state of the monitored environment. Factors such as UAV trajectory planning, scheduling of data collection visits, energy constraints, and the mobility of both UAVs and nodes significantly impact AoI performance. Therefore, advanced methods, such as reinforcement learning (RL) and deep reinforcement learning (DRL)-based algorithms, have been increasingly adopted to dynamically optimise UAV operations and minimise AoI across large and complex network deployments.

1.1.2 Deep Reinforcement Learning Applications

Deep reinforcement learning (DRL) involves the combination of deep learning and reinforcement learning. It has quickly evolved to one of the most popular machine learning domains in the last few years. DRL has proven to be able to solve a wide variety of complex decision-making problems that machines could not solve in the past. Thus, it opens up applications in domains such as smart grids, robotics, healthcare, and finance, among others [9]. For these reasons, DRL has increasingly gained popularity (see Fig. 1.1), especially due to its potential application and adoption in various fields.

DRL has gained prominence in the research on AoI minimisation in UAV-assisted IoT architectures [10]. It is effective in optimising the performance of different UAV-IoT environments, ranging from obstacle-free terrestrial settings to complex, unknown flight environments with uncertain channel conditions. In UAV-IoT applications, DRL is applied for UAV navigation for AoI minimisation, power control, energy optimisation, resource allocation, altitude control, and trajectory planning. Other key objectives include load balancing, localisation, multi-UAV control, sensor node (SN) and UAV task allocation and scheduling, UAV speed and velocity control,

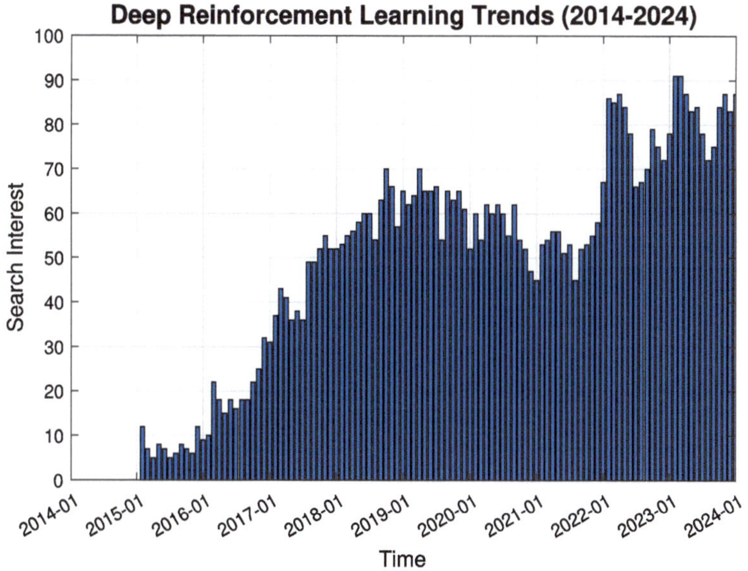

Fig. 1.1 A decade evolution of "deep reinforcement learning" Search Interest on Google Trends

real-time trajectory planning, network resource management, target tracking, SN-UAV association, energy allocation and optimisation of sensor nodes, UAV service time optimisation, as well as collection point selection and optimisation.

The above objectives can be achieved by properly mapping the desired problem into the Markov Decision Processes (MDPs) framework and deploying a suitable DRL algorithm.

1.2 UAV-Assisted WSN/IoT Architecture and Assumptions

A UAV-assisted sensor and IoT network typically involves several interconnected components such as a platform, payload, various sensors, storage, connectivity components, a flight control system, and a ground control station.

UAVs collect data from sensor nodes (SNs) or IoT devices that sample data from their environments. The uploaded data is relayed to a base station or central controller. UAVs collect sensed data via air-to-ground communication links by hovering above SNs or IoT devices. Data can be gathered using single or multiple UAVs, with sensors arranged in clustered or randomly dispersed fashions [11] as illustrated in Fig. 1.2. It is also possible to have a mixed architecture with both randomly dispersed nodes and clustered nodes. In some cases, special collection points (CPs) in such terrestrial networks become pivotal locations where UAVs hover and collect aggregated data,

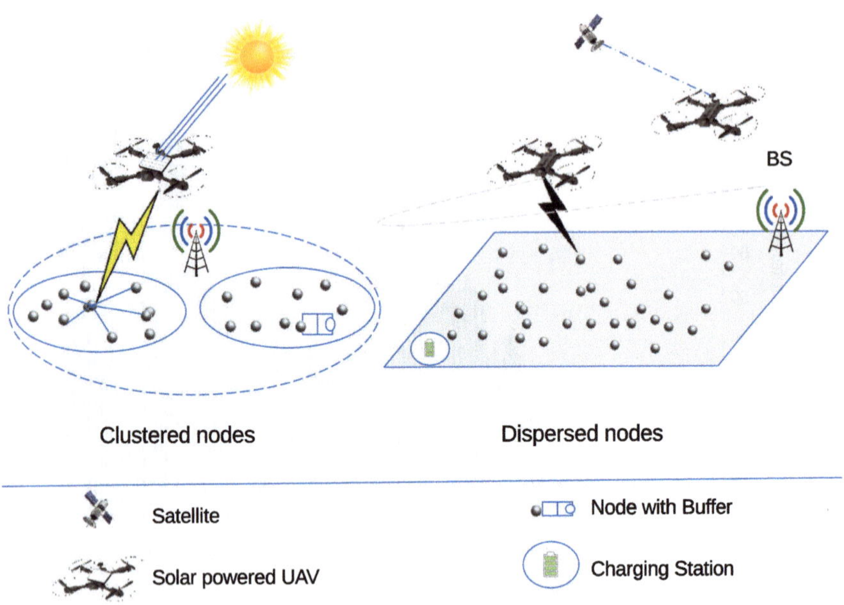

Fig. 1.2 Illustration of a UAV-assisted IoT system featuring both clustered and randomly deployed sensors

saving energy and reducing flight time. In addition, certain applications require sensor nodes with buffers to store packets, charging stations (CSs) could also be deployed to charge the UAVs, and in other cases, the UAV may be charged using renewable energy, such as solar energy. Sensor nodes may also be powered via energy harvesting, where the UAV can function as a potential energy source, providing energy for the sensor nodes. Some environments are quite complex, involving nodes with heterogeneous quality of service or time-related requirements, while others may include obstacles. In the case of obstacles, UAVs carrying reconfigurable intelligent surfaces (RIS) can be a potential solution as they help to improve the quality of signal reception. A satellite could also be involved in the communication due to its higher coverage range, as presented in [12]. In this case, the UAV can take advantage of the satellite to send data to a destination further away.

Assumptions relevant to the modeling and study of AoI-based UAV-IoT applications are numerous, as discussed in [11]. Some of them are time-related, such as a constant transmission time for data transfer from SN to UAV and a negligible sampling time by the sensor or IoT nodes. Similarly, time is commonly assumed to be slotted, and UAVs could schedule only one IoT device in a time slot using Orthogonal Frequency-Division Multiple Access (OFDMA) or Time Division Multiple Access (TDMA) protocols. Each status update collected from the IoT devices is timestamped to accurately capture its generation time. In addition, IoT devices can be selected randomly for uploading sampled data to the UAV, which is generally not time-efficient. Rather, data transmission based on AoI is a more prominent assumption. A newly arrived sample at the SN could replace an older status-update packet, or they can all be stored in a buffer of finite size before collection by the UAV. The network can be large in size and dense, and the target area for data collection can be partitioned to facilitate ease of modeling. Nodes can also be prioritised based on application requirements.

Base stations can be positioned anywhere in the network, such as in the middle or at the edge, whereas sensor nodes are deterministically positioned or randomly distributed. In many cases, the channel is unpredictable due to environmental conditions, such as obstacles and uncertain channels. The aforementioned assumptions are mostly related to the terrestrial sensor and IoT nodes.

As for UAVs, depending on the size of the network, a single UAV or multiple UAVs are deployed. These UAVs fly around the network at fixed or variable altitudes, and this could be either a single-trip flight or multiple trips. The propulsion energy consumed by the UAV is significant and thus worthy of consideration, as the UAVs have a finite battery size requiring constant charging. In many cases, the mission needs to be properly planned for completion before the UAV battery is depleted. The UAVs move at either fixed speeds or variable speeds and should have a speed limit.

1.3 UAV-Assisted Sensor and IoT Applications

Wireless sensor networks (WSNs) and UAVs are emerging as state-of-the-art solutions for large-scale monitoring of water, air, soil, and other physical characteristics in various domains. This includes environmental monitoring, agriculture, emergency situations, and security. Monitoring applications are perhaps the most widely studied application for UAV-assisted IoT. Several UAV-assisted sensor and IoT applications are discussed in the subsequent subsections.

1.3.1 Industrial and Structural Monitoring

UAV-assisted IoTs can play a vital role in industrial and structural monitoring applications, which have the potential to save costs by preventing damage to industrial equipment. For instance, in large industries, sensor nodes monitoring industrial equipment can be visited by the UAVs to inspect the status of such equipment through the collection of status updates. This technology has been utilised for oil and gas pipeline monitoring to avoid leaks and spills, thus increasing environmental safety and speeding up emergency responses [13]. In the same vein, this improves the reliability [14] of such systems.

In the field of infrastructure and bridge health monitoring, the integration of UAVs and sensor networks offers significant advantages. As buildings and bridges are continuously exposed to various forms of environmental and structural degradation, timely maintenance is critical to extending their lifespan and ensuring operational safety. UAVs, equipped with high-resolution imaging capabilities, can enhance the efficiency and sustainability of structural inspections by providing detailed visual data that supports early detection of damage and more accurate maintenance planning [15, 16]. Specifically, UAVs are increasingly deployed to monitor critical infrastructure such as bridges, buildings, and power lines, enabling the identification of potential risks and structural deficiencies before they escalate into major failures [17]. This proactive monitoring approach not only improves the overall reliability, lifespan, and safety of infrastructure asset, but also significantly reduces maintenance costs by enabling targeted interventions. Furthermore, the UAV-IoT systems can also contribute to the development of digital twins for large-scale industrial facilities, for the optimisation of predictive maintenance, operational performance, and data-driven facility management [18].

UAVs can also be utilised in smart factories to monitor the inventory in warehouses by scanning the shelves' content and providing real-time data to manage the warehouses and supply. The UAVs could also monitor the production lines and provide statistics about productivity. They could also be used to report any faults or equipment failures, thus facilitating the maintenance process and preventing accidents. The drones can be used to access the dangerous areas in factories, enhancing

Fig. 1.3 Illustration showing the use of UAVs for monitoring applications in industrial environments (*Source* https://pikwizard.com/photo/industrial-factory-silhouette-against-sky-at-sunset/ba8c0d74938cb252cbc3a04dd2ea6592/ licensed under CC0 (Creative Commons Zero), Free for personal and commercial use)

productivity and safety. Furthermore, drones can help in logistics and delivery inside factories by transporting small objects, which speeds up the production process [19].

Figure 1.3 illustrates the use of drones in an industrial environment such as a refinery, where the UAVs can be used for inspection and detection of faults within the industrial system.

1.3.2 Environmental Monitoring

In environmental monitoring applications, UAVs offer cost-effective monitoring solutions that may not be offered by satellites. IoT devices monitor physical phenomena like temperature, pollution, and humidity levels [20], which can be utilised in several environmental monitoring applications such as air and water quality, agriculture, and wildlife monitoring [21]. All these applications are appealing due to their cost-effectiveness.

IoT systems assisted by UAVs are particularly effective for air quality monitoring and pollution detection [22, 23]. The IoT sensors collect information about pollutants such as carbon dioxide (CO_2) and nitrogen oxides (NO_x) in a specific geographical area. The UAVs then transfer this information to a central processing unit to be anal-

Fig. 1.4 Illustration showing the use of UAVs in environmental monitoring applications (*Source* https://commons.wikimedia.org/wiki/File:Great_nature_landscape_Pakistan.jpg under Creative Commons Attribution-Share Alike 4.0 International)

ysed. The results of the analysis help to determine the pollution levels in that specific area based on the collected data. UAV-assisted IoT can help to capture greenhouse gas emissions [24] and facilitate timely data collection in remote areas. IoT devices visited by UAVs enable early wildfire detection, thereby safeguarding forests and air quality [25].

In marine environments, UAVs support intelligent ocean control systems by performing computational and data relay tasks, thereby complementing or offloading functions typically handled by offshore base stations. This enables flexible and efficient monitoring of the marine environment in various aspects, such as water quality monitoring and material distribution [26]. The UAVs could help in other marine aspects such as search and rescue operations, which helps accelerate access to missing persons in vast marine areas [27]. The marine coastal zones and wildlife can be easily monitored and tracked using UAV-IoT systems without disturbing the animals' natural habitat [28].

A typical illustration of UAVs deployed for environmental monitoring is provided in Fig. 1.4, showing that a UAV can provide information about the environmental conditions in a remote area.

1.3.3 Precision Agriculture

UAV-assisted IoT also brings huge benefits in agricultural applications [29–34], to support precision farming [35] and increase productivity. UAV-assisted IoT solutions monitor crops [36, 37], animals [38], pests [39], diseases, wildlife, and water quality.

Fig. 1.5 A typical illustration of a UAVs deployed for monitoring in an agricultural field (*Source* https://commons.wikimedia.org/wiki/File:Dronak_nekazaritzan.jpg under Creative Commons Attribution-Share Alike 4.0)

A typical illustration showing a UAV in an agricultural field is shown in Fig. 1.5 where sensors can be deployed on the field to collect environmental data that affect the plants, such as temperature, humidity, and soil moisture, and then analyse this data using artificial intelligence (AI) methods or big data to make insightful decisions. UAVs are used to collect the data from sensors and process it or transfer the data to the base stations to be processed [40]. The utilisation of UAV-assisted IoT systems in agriculture increases the precision and efficiency of farming by providing more intelligent solutions serving different aspects of farming [41]. These solutions include weed detection and management [42, 43] plant growth and health monitoring [44, 45], irrigation management [46], and precision spraying of seeds and fertilizer [47]. Implementing these solutions in the agricultural field brings several advantages, such as productivity increment, cost savings, early detection of diseases, and resource optimisation.

Generally, the applications of UAVs in smart farming can be summarised into several categories [33], some of which include the use of UAVs for timely sensing of different parameters, the creation of 2D or 3D maps of agricultural fields, the detection of weed infestations, and the facilitation of efficient seed and seedling planting. Additionally, UAVs can effectively and quickly spray pesticides or fertilizers, outperforming speed sprayers and wide-area sprayers. UAVs are also used to forecast and control various parameters via sensors mounted on them.

1.3.4 Smart Cities

UAVs, or drones, play a major role in developing smart solutions for modern cities, facilitating improved urban planning and sustainable advancements such as environmental planning, pollution control, and conservation efforts. For instance, in energy management, UAVs can collect data on residential energy consumption that is useful for scheduling energy provisions. UAVs can also be used for electricity restoration by monitoring and reporting the faults, which helps to enhance the emergency response and repairs in power systems [48].

Drones can also be deployed for crowd surveillance and public safety applications, as they offer cost-effective security solutions to assist investigations by law enforcement agencies. The UAVs here act as a flying eye to report any suspicious person, which requires a continuous high-resolution video transmission [49].

In smart transportation systems, the UAV-assisted IoT systems could help in traffic monitoring and management, route optimisation, and goods delivery, which specifically help to reduce congestion and emissions [50]. Furthermore, the UAVs could help in developing fully automated transportation systems to enhance the system's reliability and to improve safety and security. The UAVs here can act as a flying accident report agent, roadside unit, or police eye [51]. In these scenarios, the UAVs can fly and find the accident location and provide the rescue team with a detailed report about the accident, which helps rescuers to make appropriate decisions. Figure 1.6 illustrates the deployment of drones in cities where they can be used for diverse use cases as described in this section.

Fig. 1.6 Illustration of UAVs deployment in a typical urban environment for smart city applications (*Source* https://commons.wikimedia.org/wiki/File:Drone_view_of_a_city_(Unsplash).jpg under Creative Commons CC0 1.0 Universal Public Domain Dedication)

1.3.5 Disaster Management

Besides the infrastructure monitoring and inspection that help to avoid disasters, UAVs can also provide rapid solutions needed in disaster scenarios such as earthquakes, floods, tsunamis, volcanoes, rockfalls, and forest fires [52]. In such scenarios, UAVs and IoT devices can detect relevant information, like missing people and animals, necessary for the effective management of natural disasters.

UAVs help to improve search and rescue operations [53], by collecting data from disaster areas to provide first responders with information, such as affected area boundaries and blocked routes to speed up the rescue and search operations [54]. The UAVs can also help to detect victims in open areas like deserts and seas by providing the search team with high-resolution images and precise location of the areas that are hard to reach.

In natural disaster scenarios, UAVs can be used for logistics and resource delivery as they can easily reach the blocked areas. The UAVs can help in damage assessment operations by providing high-resolution images of the affected areas, which help to quantify the damage size and determine the required support. One essential function of UAVs during natural disasters is communication support. The UAVs act as base stations to provide a connection to the affected areas, especially when the communication infrastructure experiences huge damage. UAVs can also help in mapping and surveying the affected areas, which helps in improving the disaster recovery planning and management process [55]. Figure 1.7 illustrates two UAVs deployed in a firefighting mission. In such missions, UAVs are deployed for situational awareness, fire surveillance, or to put out fires.

Fig. 1.7 Illustration showing the use of UAVs in disaster response scenarios (*Source* https://www.flickr.com/photos/blmoregon/52243369246/in/album-72157651154979602 under https://creativecommons.org/licenses/by/2.0/)

UAVs could also help in developing early warning systems to avoid disasters. The UAVs can be deployed with the required devices, such as a global navigation satellite system (GNSS), to collect data from landslides that are hard to reach by humans. This technology helps to improve the landslide monitoring and early warning systems [56].

1.4 Age of Information (AoI)

AoI parameter captures the freshness of information from the receiver's perspective, considering the time elapsed since the last status update was generated at the source. This means that the smaller the value of AoI, the fresher the data is [57, 58]. Therefore, it is desired to keep the AoI as low as possible or at least keep it less than a specific threshold to ensure the information freshness.

To illustrate the concept of AoI, consider a scenario where a UAV hovers above sensors sequentially or collects data from designated stopping points before offloading to a base station or data center. When the UAV hovers over each sensor, the AoI at the base station starts and increases until it receives data from the UAV. To illustrate this further, assume data is sampled by a sensor at time τ_1 and received at the destination at τ_2. The AoI A_{τ_2} at time τ_2 is $\tau_2 - \tau_1$. In general, the AoI for a given update can be expressed as $\Delta_{a,b}$ such that $\Delta_{a,b} = \tau_b - \tau_a$ where τ_a is the sampling time and τ_b is the receiving time of that update. As shown in Fig. 1.8, the AoI A_{τ_b} for a typical UAV at a given time τ_b is $A_0 + \Delta_{a,b}$ where A_0 is the initial AoI. For example, the AoI at time τ_4 is $A_0 + \Delta_{1,4}$ where $\Delta_{1,4} = \tau_4 - \tau_1$ as this update was sampled at time τ_1 and received at time τ_4.

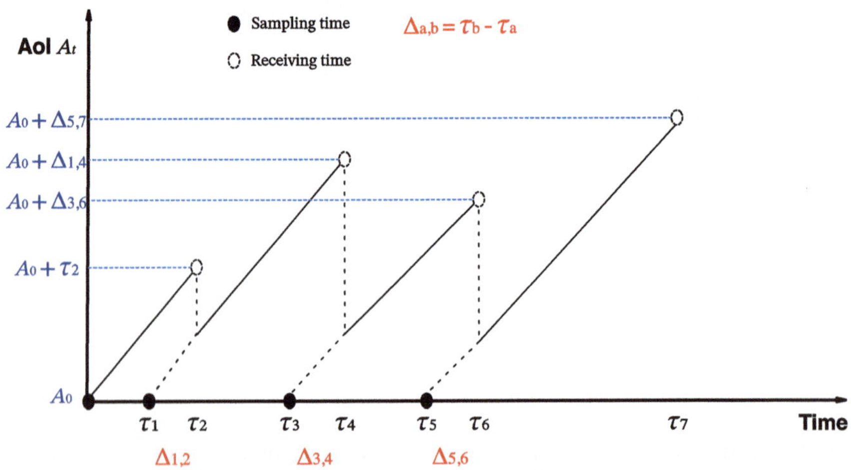

Fig. 1.8 Illustration of the Age of Information concept for a typical UAV system

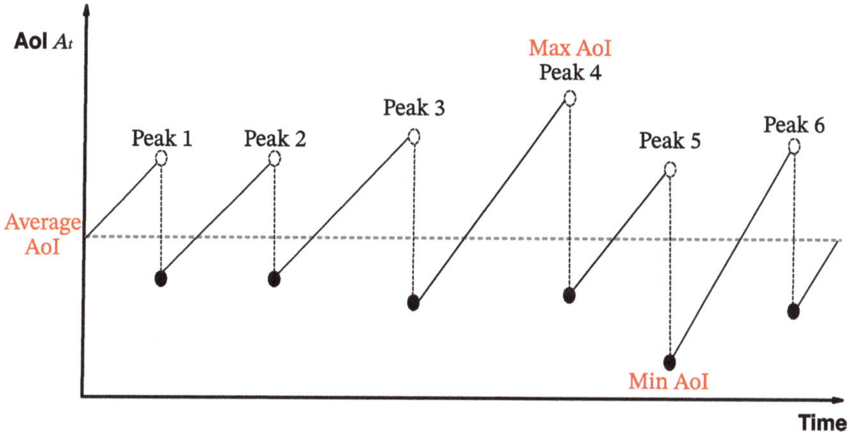

Fig. 1.9 Illustration showing various AoI metrics

AoI has been studied in different forms or metrics across diverse architectural setups. These metrics are illustrated in Fig. 1.9, and listed below:

1. **Average AoI** [58–62]: It measures the average AoI A_t in the whole system within a specific time range [63]. Average AoI incorporates factors like data collection time, UAV flight time [64], and energy transfer time [65].
2. **Total and average peak AoI** [59]: The peak here refers to the highest value of AoI for a given update. The average peak AoI is the average of all peaks (peak 1 to peak 6 in Fig. 1.9) in age evolution for any terminal, while total network peak age is the weighted average of the age of every terminal [59]. The peak AoI is used in some applications where the AoI must be less than a specific threshold with some probability [63].
3. **Maximum and minimum AoI** [66]: This metric refers to the highest and the lowest value of AoI achieved by different algorithms across different trajectories to identify the optimal one.
4. **Weighted sum AoI, and normalised weighted AoI (NWAoI)** [67]: This metric is used when the system includes different processes, each with different importance. The normalised weight is used to vary the importance of each process, which helps to find the optimal balance in AoI for all processes.

In general, the freshness of information is affected by several factors, which can be listed as follows [63]:

1. **Timeliness**: Since the AoI measures information freshness, the timeliness of data collection is crucial, especially for time-sensitive applications.
2. **Refresh rate**: How frequently the data is updated is a crucial factor in ensuring information freshness. More frequent updates lead to better data freshness.
3. **Transmission delay**: The speed at which data is transmitted between the source and destination significantly affects the AoI. Transmission delays can either

increase or decrease the AoI. Adopting efficient transmission protocols can notably reduce the AoI.

4. **Resource management**: The efficient allocation of resources, such as bandwidth and energy, can substantially improve the AoI.
5. **Queue management**: In some systems, data must wait in a queue before transmission. Efficient management of these queues is essential to ensure the freshness of critical data.
6. **Channel quality**: A strong signal between the sender and receiver ensures timely data transmission, which helps improve information freshness.

While AoI measures how old the information is, assuming that fresher is better, another key metric is the Value of Information (VoI). VoI states that freshness alone is not enough, but also considers how critical the information is to decision-making. VoI measures the importance degree of a received update, which helps to measure the cost of having old information at the destination [68]. The VoI of a given packet is the highest if it represents the latest update of important information, and this value starts to reduce once the packet is generated [63]. The VoI is used to maintain the Quality of Service (QoS) of some applications in which the optimisation of AoI does not improve the system performance, as the old information sometimes has a higher VoI compared to the latest updates. Therefore, the different updates from different sources can have various VoI regardless of their freshness [69]. However, the VoI metric is outside the scope of discussion of this book.

1.4.1 Factors Affecting Timely Data Reception in UAV-IoT Systems

Timely data transfer and reception are critical requirements in several applications such as surveillance and disaster response. In these time-sensitive scenarios, meeting strict information deadlines is essential to preserve the value of sensed data and ensure the success of UAV-assisted missions [70]. Therefore, AoI emerged as a vital metric that quantifies the freshness of received data and directly impacts decision-making processes.

However, many real-world applications face inherent challenges that hinder efficient data collection, including long transmission distances, limited battery capacities of sensors and IoT devices, and unreliable or intermittent communication links [71]. Factors such as sensor node distribution, the distance of the UAV to both the source and destination, UAV speed and altitude, and the size of the sensed data can significantly impact the AoI in UAV-IoT systems. Based on our prior studies, some of the key factors affecting timely data reception in UAV-IoT systems are depicted in Fig. 1.10, and each factor is discussed below [72].

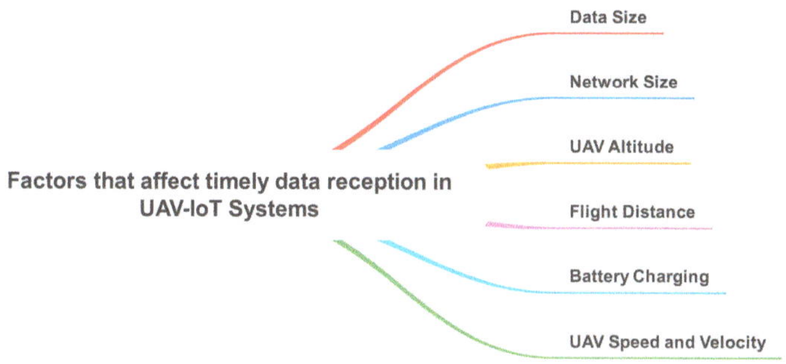

Fig. 1.10 Factors affecting timely data reception in UAV-IoT systems

1.4.1.1 Network Size

A UAV would require more time to visit all sensor nodes or collection points (CPs) in a large network. Employing multiple UAVs can reduce the age of information in such networks. Although the deployment of multiple UAVs can be quite challenging, especially when they must collaborate to perform the data collection task, each UAV must also collect data from each sensor node within the optimal time.

1.4.1.2 Battery Status

Extended UAV charging times at the cost of data collection can render SN data obsolete by the time the UAV reaches them. On the other hand, SNs that rely on ambient energy harvesting must be available to transmit data when the UAV arrives. For this reason, energy harvesting times must be considered and optimized when planning the UAV visit schedules for sensor nodes.

1.4.1.3 Flight Distance

The UAV's flight distance depends on network size and network architecture, which depend on node distribution and base station positioning. Ensuring the UAV's visits are optimised to ensure a prompt and frequent visit of sensors is vital to preserving data freshness.

1.4.1.4 UAV Speed and Velocity

UAV speed must be optimised to avoid visiting nodes before their data is ready for transmission, or after the data loses its freshness. Speed also affects energy consumption; higher speeds reduce travel time and could reduce AoI but increase energy use. Therefore, balancing the travelled UAV distance, speed, and velocity is essential.

In addition to speed control, trajectory optimisation plays a critical role, as different routes can yield varying levels of information freshness for the same energy expenditure.

1.4.1.5 UAV Altitude

A lot of energy is required for ascent in UAV applications. However, while higher altitudes increase energy consumption for ascent, they can concurrently mitigate the impact of signal blockages on long-distance sensor transmissions. However, transmitting over long distances reduces information quality due to attenuation, impacting UAV and SN energy sustainability.

1.4.1.6 Data Size

Larger data packets require longer UAV hovering times for data collection, potentially causing outdated information at other nodes. This implies that AoI increases with data size [65]. Data compression or node prioritisation can help ensure critical packets reach their destination promptly. Data aggregation, where nodes transmit to a cluster head, can also be beneficial, though it incurs some transmission overhead, especially with larger packets. Balancing these factors is crucial to maintaining optimal AoI.

1.4.2 Important Considerations in AoI-Aware UAV-IoT Design and Optimisation

Investigating several works on AoI-aware UAV-assisted data gathering applications [58–62, 64–67, 70, 71, 73–80], we find three significant themes affecting timely data dissemination: (1) UAV and SNs energy consumption, (2) UAV trajectory and (3) the scheduling of the sensor or IoT devices and the UAV. Furthermore, there are other aspects that should be considered when designing UAV-IoT systems, such as communication protocols, optimisation algorithms, queuing policy, and data privacy and security. These considerations are illustrated in Fig. 1.11, and discussed below.

1.4.2.1 Energy Management

Energy management in UAV-IoT systems constitutes energy replenishment, energy consumption minimisation, and energy efficiency maximisation. Techniques such as battery charging, charging scheduling, backscatter communication, energy trading, energy harvesting, and wireless power transfer (WPT) are employed to address these challenges.

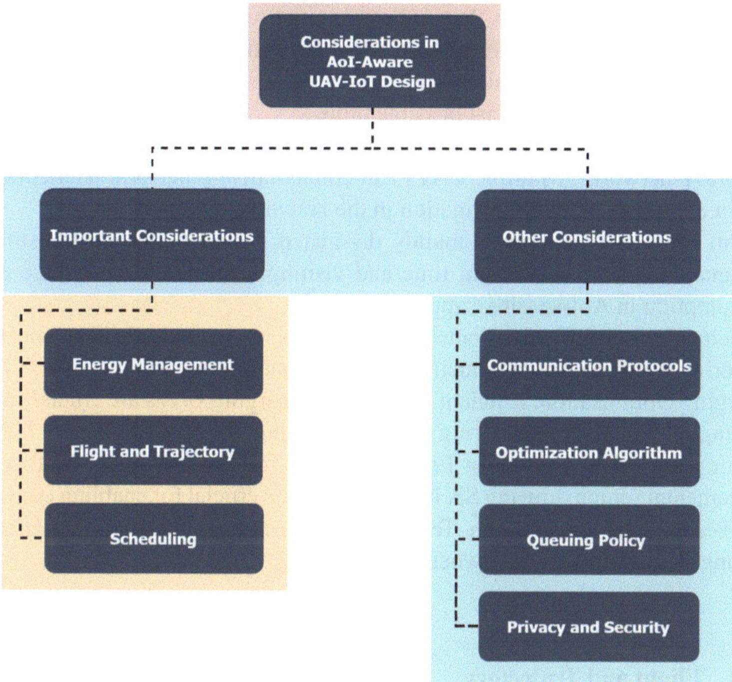

Fig. 1.11 Key considerations in AoI-aware UAV-IoT system design and Optimisation

The energy supply constraints of IoT devices make it essential to design a reliable power supply and energy-efficient data communication schemes for efficient IoT device communication. Despite their energy constraints, UAVs offer a promising solution to this challenge [81]. Nevertheless, UAVs also have energy constraints that necessitate proper scheduling of their flights. Efficient energy management prolongs the operation of both UAVs and sensor nodes [75].

It is worth noting that higher energy consumption occurs when UAVs accelerate or move quickly to collect fresh data, reducing AoI but also the UAV's lifetime. Continuous data collection applications require significant energy, necessitating optimisation [58]. Hence, scheduling policies, the choice of nodes (e.g., based on importance or priorities), and the order of visits significantly impact the UAV's energy consumption during data gathering. Several factors contributing to this energy consumption include the following [72]:

- Number of Sensor Nodes Visited: The number of IoT devices a UAV visits in a time round impacts AoI and should be optimised. In large-scale networks, using collection points can reduce the path traversed in the UAV trajectory, reducing the need to visit all sensor nodes individually, thus saving energy. Similarly, sensor nodes can also form clusters, with cluster heads uploading data to the UAV, reducing the energy burden on individual nodes.

- Energy Levels: Continuous data collection may require trading off energy consumption for minimal AoI. Ensuring UAVs have sufficient energy levels via energy supplements during continuous data collection missions is crucial for sustaining their missions within the least possible time.
- UAV Cooperation: Cooperative deployment of UAVs is advantageous in large-scale IoT networks. Multiple UAVs can collaboratively perform data collection, thus reducing the age of information in the network.
- Flight Time: UAV energy is mainly dissipated during flight and transmission. Optimal design of UAV flight time and visiting sequence helps reduce energy consumption in AoI-sensitive applications.
- Velocity Control: Effective control of UAV velocity and altitude helps meet data-gathering deadlines while reducing energy consumption.
- Charging Optimisation: Efficient methods for charging UAVs are crucial for minimising energy dissipated during visits to charging stations (CSs).
- Routing and MAC protocols: Energy-efficient routing and MAC protocols are fundamental for maximising SN lifetime, which is crucial for enabling continuous data sensing and transmission. Techniques like wake-up radio can save energy by waking sleeping nodes only when needed.

1.4.2.2 Flight and Trajectory

Efficient UAV trajectory optimisation is crucial for guaranteeing data freshness because UAV flight also impacts the AoI and, if not well managed, can lead to packet expiry. Several approaches to reduce flight time and save energy include [72]:

- Stopping Points: UAVs can stop at specific points within the network for sensor nodes to upload data before flying to another location, thus reducing flight time.
- Cluster Heads: Sensor nodes can transfer sensed information to CHs, which then transmit data to the UAV, which reduces flight time. In large-scale networks, optimising or reducing the number of CHs can achieve an age-optimal UAV trajectory.

1.4.2.3 Scheduling

Proper scheduling determines which sensor nodes sleep, and transmit data, and which UAV receives data from sensor nodes and in what order. Scheduling involves flight, hovering, transmission order and times, as well as sensor node uplink transmissions. Efficient scheduling policies can ensure timely data transmission, reducing AoI. Scheduling could also involve packet allocation, where some applications involve a uniform packet size while others require larger packets for specific nodes. Similarly, bandwidth allocation helps to ensure quality and timely data transmissions. Timely scheduling of IoT nodes is crucial to avoid longer waiting times and higher accumulated AoI. Binary indicators in mathematical models can be used to specify whether a sensor node is uploading to the UAV, which is useful for network optimisation. The sensor node-CP association and uploading sequence significantly impact AoI, making it essential to find the optimal set of CPs and UAV trajectories. Proper scheduling can reduce AoI and improve network performance, ensuring reliable and timely data collection in UAV-assisted IoT systems.

1.4.2.4 Other Design Considerations

Several design aspects must be considered when developing AoI-aware UAV-assisted IoT systems to ensure reliability and efficiency. These aspects include the communication protocols implemented, optimisation algorithms used, queuing policies adopted, and the system's privacy and security mechanisms.

- Communication Protocols: The adopted communication protocol managing the interactions between the UAV and IoT devices is a crucial aspect that must be carefully considered to ensure smooth data transmission within specified deadlines. The protocol should effectively manage interference and other factors affecting channel quality to ensure connection reliability. Various communication protocols, such as ZigBee, NB-IoT, and Long Range (LoRa), can be used in IoT transmissions [82]. The choice of the most suitable protocol depends on factors such as energy consumption, coverage range, and delay constraints.
- Optimisation Algorithms: Based on several works published on UAV-IoT systems' optimisation [58–65], it is apparent that the selection of the optimisation algorithms in UAV-IoT systems is a critical factor affecting the system's performance. For example, using an efficient optimisation algorithm for resource allocation, such as bandwidth and power, helps avoid unforeseen consequences that could degrade performance. Moreover, employing DRL algorithms to optimise UAV trajectories and task allocation can significantly enhance system reliability by reducing data collection delay, AoI, and energy consumption. Scalability in optimisation algorithms is another key design aspect to consider, especially when adding more IoT devices to the network, as it helps manage increased data traffic effectively.
- Queuing Policy: AoI is heavily influenced by the queuing policy adopted in a UAV-IoT system, as it determines the delay experienced by each packet [63]. For instance, if a First-In-First-Out (FIFO) policy is used, the most recently generated packet may experience a longer delay since it must wait for earlier packets to be transmitted. In contrast, a Last-In-First-Out (LIFO) policy allows the most recent packet to be transmitted first. While LIFO may seem more appropriate for AoI-aware systems, FIFO is also used in some cases where the system values the information carried by most packets in the buffer while maintaining data freshness. Therefore, the choice of queuing policy should be carefully made based on the application's AoI requirements.
- Privacy and Security: UAV-assisted IoT systems rely on wireless networks, which can be vulnerable to attacks and eavesdropping, potentially compromising data integrity and access authentication. Ensuring data privacy and integrity is essen-

tial for maintaining system reliability and efficiency. The data collected by UAVs must be protected against tampering to prevent misleading decisions. Thus, considering privacy and security aspects when designing UAV-assisted IoT systems is crucial. This can be achieved by implementing various procedures, including strong encryption methods to protect transmitted data, secure communication protocols to maintain authorised access and data integrity, and appropriate access control mechanisms to ensure that UAVs and IoT devices are only accessible by authorised personnel [83]. Compromised security, leading to data loss or interception, would indefinitely increase the AoI as the data would fail to reach its destination.

1.5 Summary

This chapter introduces the UAV-assisted paradigm for sensor networks and Internet of Things applications, and Age of Information, which measures data timeliness from source to destination. AoI is important when UAVs collect data from sensor nodes or IoT devices. DRL enhances UAV performance to minimise AoI by optimising energy, trajectory planning, resource allocation, or scheduling. The chapter overviews UAV-assisted WSN/IoT architectures where UAVs collect data for relay to base stations or central controllers. The assumptions and applications of such architectures are also presented. Some of the discussed assumptions on the networks include network size and density, node priority, position of base station, time-related issues, and channel obstructions. Additionally, a single UAV or multiple UAVs can be deployed with different altitudes, speeds, data-collection trips, and battery conditions. Some of the critical applications in which UAV-assisted sensor and IoT applications have been deployed, including monitoring, precision agriculture, smart cities, and disaster management, are also discussed.

This chapter also discusses the AoI concept and describes its variants, including the widely studied average AoI, along with total and average peak AoI, maximum and minimum AoI, weighted sum AoI, and normalized weighted AoI. The factors that contributed to the AoI performance in traditional wireless networks are also presented, which include timeliness, transmission delay, resource management, queue management, and channel quality. This chapter then discusses the factors affecting AoI specific to UAV-IoT systems, such as network size, battery charging, flight distance, UAV speed and velocity, UAV altitude, and data size that could affect the timeliness of the UAV's data collection process.

Subsequently, this chapter delves into key technical considerations for AoI in UAV-assisted WSN/IoT networks. These include the crucial aspect of energy management, which requires efficient techniques for energy replenishment, energy consumption minimization, and energy efficiency maximization. Further considerations encompass efficient flight and trajectory optimisation, robust scheduling, and effective queuing policies. Moreover, the selection of appropriate optimisation algorithms, communication protocols, and integrated privacy and security measures also plays a vital role. This chapter also briefly introduces another information-based metric called Value of Information (VoI), though not commonly studied. While AoI is used to measure the freshness of data received by the recipient, VoI is used to measure the importance of a received data update.

References

1. B. Turan, S.N. Turhan, Ö. Pınarer, E. Bozkaya, Simulation of unmanned aerial vehicle for mapping, in *29th Signal Processing and Communications Applications Conference (SIU)*, (IEEE, 2021), pp. 1–4
2. M. Mozaffari, W. Saad, M. Bennis, Y.-H. Nam, M. Debbah, A tutorial on UAVs for wireless networks: applications, challenges, and open problems. IEEE Commun. Surv. & Tutor. **21**(3), 2334–2360 (2019)
3. X. Wang, L. Duan, Economic analysis of unmanned aerial vehicle (UAV) provided mobile services. IEEE Trans. Mob. Comput. **20**(5), 1804–1816 (2020)
4. J. Fraden, J. King, *Handbook of Modern Sensors: Physics, Designs, and Applications*, vol. 3 (Springer, 2004)
5. M. Coccia, S. Roshani, M. Mosleh, Scientific developments and new technological trajectories in sensor research. Sensors **21**(23), 7803 (2021)
6. N.P. Mahalik, *Sensor Networks and Configuration* (Springer, 2007)
7. S. Kumar, P. Tiwari, M. Zymbler, Internet of things is a revolutionary approach for future technology enhancement: a review. J. Big Data **6**(1), 1–21 (2019)
8. P.P. Ray, A survey on internet of things architectures. J. King Saud Univ. Comput. Inf. Sci. **30**(3), 291–319 (2018)
9. V. François-Lavet, P. Henderson, R. Islam, M.G. Bellemare, J. Pineau et al., An introduction to deep reinforcement learning. Found. Trends® Mach. Learn. **11**(3–4), 219–354 (2018)
10. U.A. Bukar, M.S. Sayeed, S.F.A. Razak, S. Yogarayan, O.A. Amodu, An exploratory bibliometric analysis of the literature on the age of information-aware unmanned aerial vehicles aided communication. Informatica **47**(7)
11. O.A. Amodu, U.A. Bukar, R.A.R. Mahmood, C. Jarray, M. Othman, Age of information minimization in uav-aided data collection for WSN and IoT applications: a systematic review. J. Netw. Comput. Appl. 103652 (2023)
12. C. Zhou, H. He, P. Yang, F. Lyu, W. Wu, N. Cheng, X. Shen, Deep RL-based trajectory planning for AoI minimization in UAV-assisted IoT, in *2019 11th International Conference on Wireless Communications and Signal Processing (WCSP)*, (IEEE, 2019), pp. 1–6
13. S.N. Karam, K. Bilal, J. Shuja, L.U. Khan, M. Bilal, M.K. Khan, Intelligent IoT-and UAV-assisted architecture for pipeline monitoring in OGI. IT Prof. **26**(3), 46–54 (2024)
14. M. Salhaoui, A. Guerrero-González, M. Arioua, F.J. Ortiz, A. El Oualkadi, C.L. Torregrosa, Smart industrial iot monitoring and control system based on UAV and cloud computing applied to a concrete plant. Sensors **19**(15), 3316 (2019)
15. H.I. Moud, M. Gheisari, Coupling wireless sensor networks and unmanned aerial vehicles in bridge health monitoring systems, in *ISARC 2016–33rd International Symposium on Automation and Robotics in Construction*, pp. 267–273 (2016)
16. F. Ahmed, J. Mohanta, A. Keshari, P.S. Yadav, Recent advances in unmanned aerial vehicles: a review. Arab. J. Sci. Eng. **47**(7), 7963–7984 (2022)
17. N. Abbas, Z. Abbas, X. Liu, S.S. Khan, E.D. Foster, S. Larkin, A survey: future smart cities based on advance control of unmanned aerial vehicles (UAVs). Appl. Sci. **13**(17), 9881 (2023)
18. Y. Sun, H. Fesenko, V. Kharchenko, L. Zhong, I. Kliushnikov, O. Illiashenko, O. Morozova, A. Sachenko, UAV and IoT-based systems for the monitoring of industrial facilities using digital twins: methodology, reliability models, and application. Sensors **22**(17), 6444 (2022)

19. P.M. Hell, P.J. Varga, Drone systems for factory security and surveillance. Interdiscip. Descr. Complex Syst. INDECS **17**(3-A), 458–467 (2019)
20. M.A. Abd-Elmagid, N. Pappas, H.S. Dhillon, On the role of age of information in the internet of things. IEEE Commun. Mag. **57**(12), 72–77 (2019)
21. S. Manfreda, M.F. McCabe, P.E. Miller, R. Lucas, V. Pajuelo Madrigal, G. Mallinis, E. Ben Dor, D. Helman, L. Estes, G. Ciraolo et al., On the use of unmanned aerial systems for environmental monitoring. Remote. Sens. **10**(4), 641 (2018)
22. J.-I. Hernández-Vega, E.R. Varela, N.H. Romero, C. Hernández-Santos, J.L.S. Cuevas, D.G.P. Gorham, Internet of things (iot) for monitoring air pollutants with an unmanned aerial vehicle (uav) in a smart city, in *Smart Technology* (Springer, 2018), pp. 108–120
23. V. Ninkovic, D. Vukobratovic, D. Miskovic, UAV-assisted distributed learning for environmental monitoring in rural environments, in *2024 7th International Balkan Conference on Communications and Networking (BalkanCom)*, (IEEE, 2024), pp. 296–300
24. B. Potter, G. Valentino, L. Yates, T. Benzing, A. Salman, Environmental monitoring using a drone-enabled wireless sensor network, in *2019 Systems and Information Engineering Design Symposium (SIEDS)*, (IEEE, 2019), pp. 1–6
25. M. Zhang, X. Li, Drone-enabled internet-of-things relay for environmental monitoring in remote areas without public networks. IEEE Internet Things J. **7**(8), 7648–7662 (2020)
26. M. Yan, H. Yuan, J. Xu, Y. Yu, L. Jin, Task allocation and route planning of multiple UAVs in a marine environment based on an improved particle swarm optimization algorithm. EURASIP J. Adv. Signal Process. **2021**, 1–23 (2021)
27. L. Lyu, Z. Chu, B. Lin, Joint association and power optimization for multi-UAV assisted cooperative transmission in marine IoT networks. Peer Peer Netw. Appl. **14**(5), 3307–3318 (2021)
28. R. Adade, A.M. Aibinu, B. Ekumah, J. Asaana, Unmanned aerial vehicle (UAV) applications in coastal zone management-a review. Environ. Monit. Assess. **193**, 1–12 (2021)
29. F.G. Costa, J. Ueyama, T. Braun, G. Pessin, F.S. Osório, P.A. Vargas, The use of unmanned aerial vehicles and wireless sensor network in agricultural applications, in *2012 IEEE International Geoscience and Remote Sensing Symposium*, (IEEE, 2012), pp. 5045–5048
30. M.M. Vihari, U.R. Nelakuditi, M.P. Teja, Iot based unmanned aerial vehicle system for agriculture applications, in *2018 International Conference on Smart Systems and Inventive Technology (ICSSIT)*, (IEEE, 2018), pp. 26–28
31. T. Moribe, H. Okada, K. Kobayashl, M. Katayama, Combination of a wireless sensor network and drone using infrared thermometers for smart agriculture, in *2018 15th IEEE Annual Consumer Communications & Networking Conference (CCNC)*, (IEEE, 2018), pp. 1–2
32. P. Radoglou-Grammatikis, P. Sarigiannidis, T. Lagkas, I. Moscholios, A compilation of UAV applications for precision agriculture. Comput. Netw. **172**, 107148 (2020)
33. N. Islam, M.M. Rashid, F. Pasandideh, B. Ray, S. Moore, R. Kadel, A review of applications and communication technologies for internet of things (iot) and unmanned aerial vehicle (uav) based sustainable smart farming. Sustainability **13**(4), 1821 (2021)
34. S. Punia, H. Krishna, V. Navada, A. Sajjad, Agrosquad-an iot based precision agriculture using uav and low-power soil multi-sensor, in *2021 IEEE International Conference on Electronics, Computing and Communication Technologies (CONECCT)*, (IEEE, 2021), pp. 1–6
35. P.K. Singh, A. Sharma, An intelligent WSN-UAV-based IoT framework for precision agriculture application. Comput. Electr. Eng. **100**, 107912 (2022)
36. S. Jayanthy, G. Kiruthika, G. Lakshana, M. Pragatheshwaran, Early cotton plant disease detection using drone monitoring and deep learning, in *2024 IEEE International Conference for Women in Innovation, Technology & Entrepreneurship (ICWITE)*, (IEEE, 2024), pp. 625–630
37. M. Ammad Uddin, A. Mansour, D. Le Jeune, M. Ayaz, E.-H.M. Aggoune, UAV-assisted dynamic clustering of wireless sensor networks for crop health monitoring. Sensors **18**(2), 555 (2018)
38. J. Xu, G. Solmaz, R. Rahmatizadeh, D. Turgut, L. Bölöni, Animal monitoring with unmanned aerial vehicle-aided wireless sensor networks, in *2015 IEEE 40th Conference on Local Computer Networks (LCN)*, (IEEE, 2015), pp. 125–132

39. G. Ristorto, A. Bojeri, G. Scarton, G. Giannotta, G. Guglieri, droneONtrap project-integration of IoT technologies and drones for health status and pests monitoring of orchards, in *2021 IEEE International Workshop on Metrology for Agriculture and Forestry (MetroAgriFor)*, (IEEE, 2021), pp. 12–16
40. N. Cheng, S. Wu, X. Wang, Z. Yin, C. Li, W. Chen, F. Chen, AI for UAV-assisted IoT applications: a comprehensive review. IEEE Internet Things J. **10**(16), 14438–14461 (2023)
41. D. Tsouros, S. Bibi, P. Sarigiannidis, A review on UAV-based applications for precision agriculture. Information **10**(11), 349 (2019)
42. M.D. Bah, A. Hafiane, R. Canals, Weeds detection in UAV imagery using SLIC and the hough transform, in *2017 Seventh International Conference on Image Processing Theory, Tools and Applications (IPTA)*, (IEEE, 2017), pp. 1–6
43. A. dos Santos Ferreira, D.M. Freitas, G.G. da Silva, H. Pistori, M.T. Folhes, Weed detection in soybean crops using convnets. Comput. Electron. Agric. **143**, 314–324 (2017)
44. J. Jung, M. Maeda, A. Chang, J. Landivar, J. Yeom, J. McGinty, Unmanned aerial system assisted framework for the selection of high yielding cotton genotypes. Comput. Electron. Agric. **152**, 74–81 (2018)
45. M. Kerkech, A. Hafiane, R. Canals, Deep leaning approach with colorimetric spaces and vegetation indices for vine diseases detection in UAV images. Comput. Electron. Agric. **155**, 237–243 (2018)
46. C. Albornoz, L.F. Giraldo, Trajectory design for efficient crop irrigation with a UAV, in *2017 IEEE 3rd Colombian Conference on Automatic Control (CCAC)*, (IEEE, 2017), pp. 1–6
47. X. Xue, Y. Lan, Z. Sun, C. Chang, W.C. Hoffmann, Develop an unmanned aerial vehicle based automatic aerial spraying system. Comput. Electron. Agric. **128**, 58–66 (2016)
48. X. Chen, Y. Li, G. Jia, M. Zhu, L. Li, Research on the application technology of UAV in restoring electricity in residential area, in *Eighth International Conference on Energy System, Electricity, and Power (ESEP 2023)*, vol. 13159 (SPIE, 2024), pp. 2634–2640
49. M.C. Lucic, O. Bouhamed, H. Ghazzai, A. Khanfor, Y. Massoud, Leveraging UAVs to enable dynamic and smart aerial infrastructure for its and smart cities: an overview. Drones **7**(2), 79 (2023)
50. P. Mishra, G. Singh, Unmanned aerial vehicles in sustainable smart cities, in *Sustainable Smart Cities: Enabling Technologies, Energy Trends and Potential Applications* (Springer, 2023), pp. 221–238
51. H. Menouar, I. Guvenc, K. Akkaya, A.S. Uluagac, A. Kadri, A. Tuncer, UAV-enabled intelligent transportation systems for the smart city: applications and challenges. IEEE Commun. Mag. **55**(3), 22–28 (2017)
52. M.A. Al-Mashhadani, M.M. Hamdi, A.S. Mustafa, Role and challenges of the use of UAV-aided wsn monitoring system in large-scale sectors, in *2021 3rd International Congress on Human-Computer Interaction, Optimization and Robotic Applications (HORA)*, (IEEE, 2021), pp. 1–5
53. A.H. Wheeb, R. Nordin, A. Samah, D. Kanellopoulos, Performance evaluation of standard and modified OLSR protocols for uncoordinated UAV ad-hoc networks in search and rescue environments. Electronics **12**(6), 1334 (2023)
54. W. Alawad, N.B. Halima, L. Aziz, An unmanned aerial vehicle (UAV) system for disaster and crisis management in smart cities. Electronics **12**(4), 1051 (2023)
55. I. Chandran, K. Vipin, Multi-UAV networks for disaster monitoring: challenges and opportunities from a network perspective. Drone Syst. Appl. **12**, 1–28 (2024)
56. Q. Zhang, Z. Bai, G. Huang, J. Kong, Y. Du, D. Wang, C. Jing, W. Xie, Innovative landslide disaster monitoring: unmanned aerial vehicle-deployed GNSS technology. Geomat. Nat. Hazards Risk **15**(1), 2366374 (2024)
57. N. Pappas, M.A. Abd-Elmagid, B. Zhou, W. Saad, H.S. Dhillon, *Age of Information: Foundations and Applications* (Cambridge University Press, 2023)
58. M. Sun, X. Xu, X. Qin, P. Zhang, AoI-energy-aware UAV-assisted data collection for IoT networks: a deep reinforcement learning method. IEEE Internet Things J. **8**(24), 17275–17289 (2021). https://doi.org/10.1109/JIOT.2021.3078701

59. X. Zeng, F. Ma, T. Chen, X. Chen, X. Wang, Age-optimal UAV trajectory planning for information gathering with energy constraints, in *2020 IEEE/CIC International Conference on Communications in China (ICCC)*, (IEEE, 2020), pp. 881–886
60. M.A. Abd-Elmagid, A. Ferdowsi, H.S. Dhillon, W. Saad, Deep reinforcement learning for minimizing age-of-information in UAV-assisted networks, in *2019 IEEE Global Communications Conference (GLOBECOM)*, (IEEE, 2019), pp. 1–6
61. N. Zhang, J. Liu, L. Xie, P. Tong, A deep reinforcement learning approach to energy-harvesting UAV-aided data collection, in *2020 International Conference on Wireless Communications and Signal Processing (WCSP)*, (IEEE, 2020), pp. 93–98
62. M. Yi, X. Wang, J. Liu, Y. Zhang, B. Bai, Deep reinforcement learning for fresh data collection in UAV-assisted IoT networks, in *IEEE INFOCOM 2020-IEEE Conference on Computer Communications Workshops (INFOCOM WKSHPS)*, (IEEE, 2020), pp. 716–721
63. Q. Abbas, S.A. Hassan, H.K. Qureshi, K. Dev, H. Jung, A comprehensive survey on age of information in massive IoT networks. Comput. Commun. **197**, 199–213 (2023)
64. Z. Yuhua, J. Xiangdong, W. Yuxin, C. Shengnan, A research on age of information minimization scheme of wireless sensor network assisted by UAV, in *2021 6th International Conference on Intelligent Computing and Signal Processing (ICSP)*, (IEEE, 2021), pp. 1149–1152
65. H. Hu, K. Xiong, G. Qu, Q. Ni, P. Fan, K.B. Letaief, Aoi-minimal trajectory planning and data collection in UAV-assisted wireless powered IoT networks. IEEE Internet Things J. **8**(2), 1211–1223 (2020)
66. J. Liu, X. Wang, B. Bai, H. Dai, Age-optimal trajectory planning for UAV-assisted data collection, in *IEEE INFOCOM 2018-IEEE Conference on Computer Communications Workshops (INFOCOM WKSHPS)*, (IEEE, 2018), pp. 553–558
67. A. Ferdowsi, M.A. Abd-Elmagid, W. Saad, H.S. Dhillon, Neural combinatorial deep reinforcement learning for age-optimal joint trajectory and scheduling design in UAV-assisted networks. IEEE J. Sel. Areas Commun. **39**(5), 1250–1265 (2021)
68. A. Kosta, N. Pappas, A. Ephremides, V. Angelakis, Age and value of information: non-linear age case, in *2017 IEEE International Symposium on Information Theory (ISIT)*, (IEEE, 2017), pp. 326–330
69. M. Costa, M. Codreanu, A. Ephremides, On the age of information in status update systems with packet management. IEEE Trans. Inf. Theory **62**(4), 1897–1910 (2016)
70. W. Li, L. Wang, A. Fei, Minimizing packet expiration loss with path planning in UAV-assisted data sensing. IEEE Wirel. Commun. Lett. **8**(6), 1520–1523 (2019)
71. M.A. Abd-Elmagid, H.S. Dhillon, Average peak age-of-information minimization in UAV-assisted IoT networks. IEEE Trans. Veh. Technol. **68**(2), 2003–2008 (2018)
72. O.A. Amodu, R. Nordin, C. Jarray, U.A. Bukar, R.A. Raja Mahmood, M. Othman, A survey on the design aspects and opportunities in age-aware UAV-aided data collection for sensor networks and internet of things applications. Drones **7**(4), 260 (2023)
73. C. Zhou, H. He, P. Yang, F. Lyu, W. Wu, N. Cheng, X. Shen, Deep RL-based trajectory planning for AoI minimization in UAV-assisted IoT, in *2019 11th International Conference on Wireless Communications and Signal Processing (WCSP)*, (IEEE, 2019), pp. 1–6
74. P. Tong, J. Liu, X. Wang, B. Bai, H. Dai, UAV-enabled age-optimal data collection in wireless sensor networks, in *2019 IEEE International Conference on Communications Workshops (ICC Workshops)*, (IEEE, 2019), pp. 1–6
75. A. Changizi, M.J. Emadi, Age-optimal path planning for finite-battery UAV-assisted data dissemination in IoT networks. IET Commun
76. G. Ahani, D. Yuan, Y. Zhao, Age-optimal UAV scheduling for data collection with battery recharging. IEEE Commun. Lett. **25**(4), 1254–1258 (2021). https://doi.org/10.1109/LCOMM.2020.3047909
77. J. Liu, P. Tong, X. Wang, B. Bai, H. Dai, UAV-aided data collection for information freshness in wireless sensor networks. IEEE Trans. Wirel. Commun. **20**(4), 2368–2382 (2021). https://doi.org/10.1109/TWC.2020.3041750
78. M. Samir, C. Assi, S. Sharafeddine, A. Ghrayeb, Online altitude control and scheduling policy for minimizing AoI in UAV-assisted IoT wireless networks. IEEE Trans. Mob. Comput. **21**(7), 2493–2505 (2022). https://doi.org/10.1109/TMC.2020.3042925

79. C. Mao, J. Liu, L. Xie, Multi-UAV aided data collection for age minimization in wireless sensor networks, in *2020 International Conference on Wireless Communications and Signal Processing (WCSP)*, (IEEE, 2020), pp. 80–85
80. M. Samir, M. Elhattab, C. Assi, S. Sharafeddine, A. Ghrayeb, Optimizing age of information through aerial reconfigurable intelligent surfaces: a deep reinforcement learning approach. IEEE Trans. Veh. Technol. **70**(4), 3978–3983 (2021)
81. X. Lu, W. Yang, S. Yan, Z. Li, D.W.K. Ng, Covertness and timeliness of data collection in UAV-aided wireless-powered IoT. IEEE Internet Things J
82. T. Lagkas, V. Argyriou, S. Bibi, P. Sarigiannidis, UAV IoT framework views and challenges: towards protecting drones as "Things". Sensors **18**(11), 4015 (2018)
83. J.-P. Yaacoub, H. Noura, O. Salman, A. Chehab, Security analysis of drones systems: attacks, limitations, and recommendations. Internet Things **11**, 100218 (2020)

Chapter 2
AoI-Aware UAV-IoT Modeling Using MDPs

2.1 Introduction to Markov Decision Processes (MDPs)

MDPs provide a framework for modeling sequential decision-making problems where outcomes are partly deterministic and random. MDPs involve decision-making over distinct times or intervals where an agent's actions on the environment influence immediate rewards and subsequent states through future rewards [1]. MDPs are fundamental to developing reinforcement learning-based solutions. One example of the use of reinforcement learning is for planning a UAV trajectory to minimise AoI. In this scenario, the UAV functions as the agent and learns how to fly around to collect data within the least possible time. The UAV takes specific actions in the environment, which may affect the environment and bring it to a new state. So, for instance, if it moves right or left, this constitutes a new state. The UAV gets a reward or penalty for each new state in the environment. The agent interacts with the environment in episodes. Suppose the UAV reads the state at the current time t denoted as $s(t)$. The agent uses its policy $\pi(a \mid s)$ to decide which actions (a) to take (e.g., directions to move) based on its current state. The action of the UAV is *transmitted* to the environment, and the time step advances. After performing the action $a(t)$, the environment transitions to the next state $s(t+1)$ and receives a reward $r(t+1)$. This forms a tuple: $(s(t), a(t), r(t+1), s(t+1))$, which represents a single transition or step of an MDP. Note that the tuple may also be written as (s, a, r', s') where the prime symbol indicates the next state. The environment is represented by the probability distribution $P(s', r \mid s, a)$. This makes it easier to understand and generalize the environment dynamics in real-world situations, and representing the environment in a probabilistic fashion makes it possible to handle uncertainty and imperfect information, especially in scenarios where the agent does not have full knowledge of the environment. Thus, the MDP framework is constituted by states, S, actions, A, transition probability, P (from state s to next state s' after executing an action, a), and the immediate reward, r, obtained after the agent performs the action.

O. A. Amodu et al., *Markov Decision Processes and Reinforcement Learning for Timely UAV-IoT Data Collection Applications*, Studies in Computational Intelligence 1220,
https://doi.org/10.1007/978-3-031-97011-5_2

Thus, it can be represented as a tuple (S, A, P, r) [2], especially when the discount factor is set to the default.

MDPs are either fully observable or partially observable. In fully observable MDPs, agents can fully observe the states, and there are no hidden states and controls. On the other hand, partially observable MDPs (POMDPs) involve states that cannot be fully observed by agents. Thus, agents maintain a state probability distribution based on observations and observation probabilities [3] in such cases. Figure 2.1 shows the architectures of both MDP and POMPDP. In UAV-assisted IoT frameworks, several problems solved using DRL are formulated as an MDP, which can be used as a basis to optimise the UAV trajectory, sensor node scheduling, and energy management, and determine optimal policies via modelled states, actions, and rewards [4, 5].

MDP problems can be solved using dynamic programming methods. An agent interacts with its environment and explores it for the best policy for prescribing actions that maximise the cumulative total of rewards [6]. MDPs can be extended to multi-agent MDPs with several agents that interact to achieve a common goal in a shared environment. An MDP consists of a tuple (S, A, P, R, γ), in which S is the finite set of states, A represents actions, P is the state transfer process of the environment after receiving an action, R is the reward function, and $\gamma \in [0, 1]$ is the discount factor. Mathematically, it can be expressed in the equations below:

$$P^{a}_{SS'} = P[S_{t+1} = S' | S_t = S,\ A_t = a],$$

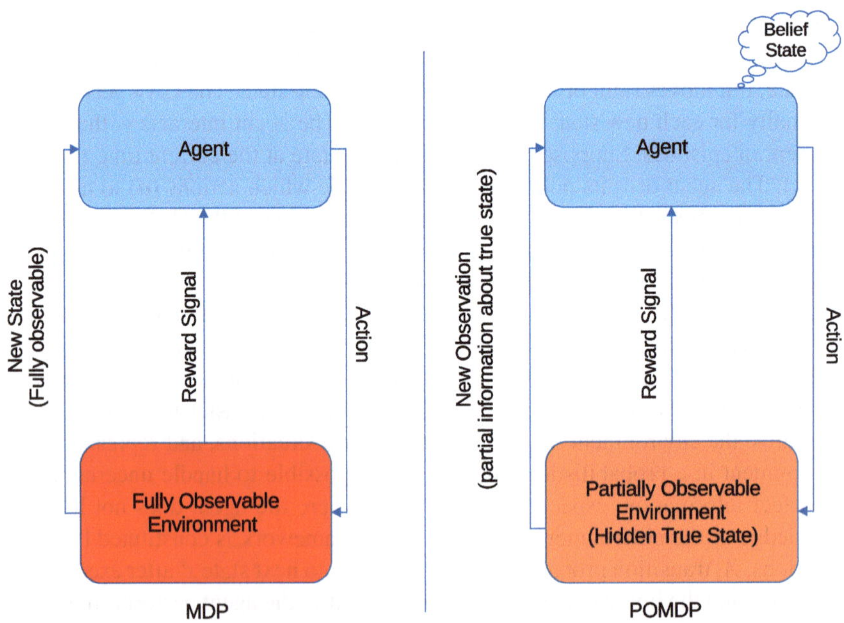

Fig. 2.1 Typical agent-environment interaction in MDP and POMDP frameworks

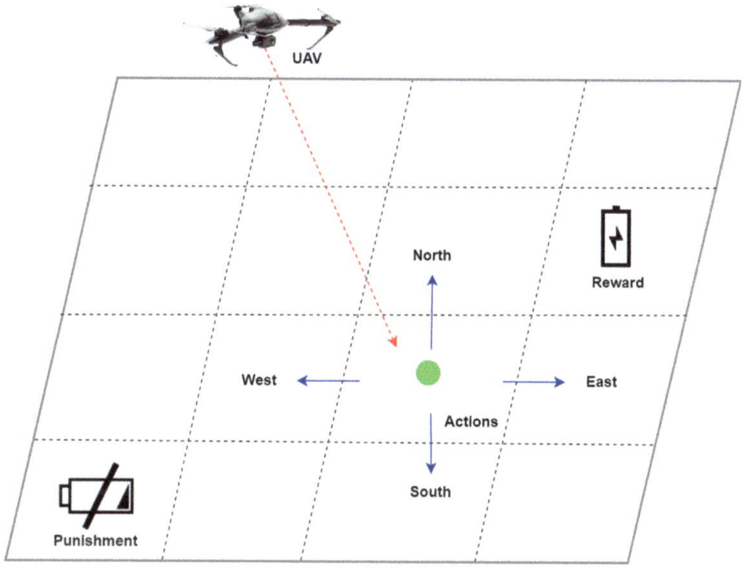

Fig. 2.2 A simple abstraction of actions on a grid, where each box represents a distinct state

$$R_S^a = E[R_{t+1}|S_t = S,\ A_t = a],$$

in which $P_{SS'}^a$ is the state transfer probability function that indicates the probability that the agent (e.g., UAV) moves to the next state after receiving the action in the current state. Also, the reward function, R_S^a, indicates the reward given back to the agent by the environment after it receives an action in the current state. It is worth noting that many of the examples presented in this chapter have assumed the partition of the deployment environment into a grid form, to facilitate abstracting actions into cardinal directions, as shown in Fig. 2.2.

2.2 Fully Observable MDP-Based Modeling and Problem Formulation

In this section, examples of MDP problem formulation are discussed based on several published studies. Common notations used in this section are provided in Table 2.1. Some of the previously discussed studies are classified as either fully or partially observable MDP problems. A fully observable environment has access to all required information to complete the target task, and most of the discussed works fall under this category. On the other hand, a partially observable environment offers only partial information, and only a few have been identified as falling

Table 2.1 Table of notations

Symbol	Description
t	Time step or time slot (in general)
i	Index for a general agent (UAV or sensor node)
u	Index for a UAV
k	Index for a sensor node or IoT device
j	Index for a virtual queue
K	Total number of sensor nodes or IoT devices
M	Total number of UAVs
S	State space
s_t	State at time t
$s_i(t)$	State of agent i at time t
$s_u(t)$	State of UAV u at time t
$s_k(t)$	State of sensor node k at time t
A	Action space
a_t	Action at time t
$a_i(t)$	Action of agent i at time t
$a_u(t)$	Action of UAV u at time t
$a_k(t)$	Action of sensor node k at time t
r_t	Reward at time t
$r_i(s_i(t), a_t)$	Reward of agent i at time t given state and action
$A_k(t)$	Age of Information of IoT device k at time t
$\mathbf{A}(t)$	Vector of AoI of all IoTDs at time t
$x_u(t)$	Horizontal coordinate of UAV u at time t
$y_u(t)$	Vertical coordinate of UAV u at time t
$v_u(t)$	Speed of UAV u at time t
$\theta_u(t)$	Flying direction of UAV u at time t
$E_k(t)$	Energy in the battery of ground node k at time t
$\eta_u(t)$	Remaining energy of UAV u at time t
$\delta_u(t)$	AoI of the state observed by UAV u at time t
$v(t)$	Flight speed of the UAV at time t (if referring to a generic UAV)
Δt_{flight}	Flying duration
$d_u(t)$	Flight distance of UAV u at time t
$\mathbf{p}_u(t)$	Position vector of UAV u at time t
d_{flight}	Flight distance (if referring to a generic flight)
η_{BS}	Average energy efficiency of the UAV-BS
Δ_{nav}	Average age for navigation optimisation
$E_{eff,u}(t)$	Energy efficiency of UAV u at time t
ϕ	Normalised energy efficiency parameter
$R^t_{\{k,i,l\}}$	Transmit rate of link k_i over sub-carrier l at time t
$P_{\{k,i,l\}}$	Transmit power of sub-carrier l for link k_i

(continued)

Table 2.1 (continued)

Symbol	Description
L_{PC}	Length of the power control action space
A_u^t	Flight action of UAV u at time t
L_{TD}	Length of the trajectory design action space
$R_{\text{sum},i}$	Total communication rate over sub-carriers assigned to link k_i
$\Psi_{\text{PC},t,k,i,l}$	Reward for power control
ω_u	AoI value for the u-th UAV
λ	Weight balancing communication rate and AoI
$\mathbf{p}_{\text{current}}^u$	Current location of UAV u
$\mathbf{p}_{\text{target}}$	Target position
η_{th}	Energy efficiency threshold
$\hat{\Delta}_b^{th}$	Maximum average age threshold
$R(a_t)$	Immediate reward at time t given action a_t
γ	Discount factor
$h_k(t)$	Index of the hexagon above which sensor node k is located at time t
$\delta(t)$	Current AoI of all IoTDs at time t
$\mathcal{N}(h_k(t))$	Set of neighbouring hexagons of $h_k(t)$
$\delta_k(t)$	AoI evolution of IoTD k at time t
$\mathbf{l}(t)$	Vector of positions of all UAVs at time t
$\mathbf{A}_U(t)$	Vector of AoI of all SNs under UAV coverage at time t
$\delta_U(t)$	Vector of battery status differences for each UAV at time t
$\mathcal{L}_u$	Sequence of UAV u's locations over a trip
r^α	Information age reward
r^c	Data collection reward
$\mathcal{S}^U$	State space for UAVs
A^U	Action space for UAVs
$\mathbf{v}_u(t)$	Direction vector of UAV u's movement at time t
θ_k	Weight denoting importance of SN k
$d_{u,k}$	Distance between UAV u and SN k
R_k	Coverage radius of sensor k
$p_{penalty}$	Energy penalty
$t_{u,final}(t)$	Time left for UAV u to reach final location at time t
$a_k(t)$	AoI of ground node k at time t
$e_k^T(t)$	Energy for data transmission of ground node k at time t
$e_k^H(t)$	Harvested energy block of ground node k at time t
$a_s(t)$	Scheduling action for ground nodes at time t
$J^\pi(s(1))$	Total expected cost over a finite horizon for policy π starting at $s(1)$
$\mathcal{T}$	Set of discrete time steps
λ_k	Importance weight of the process observed by node k
$\mathbf{p}_F$	Final location of UAV

(continued)

Table 2.1 (continued)

Symbol	Description
Δt_{slot}	Duration of a time slot
$\mathbf{o}(t)$	UAV projection on the ground at time t
$\phi(t)$	Time difference between remaining time and minimum required time
δ_p	Penalty term for incorrect moves or violations of constraints
$b(t)$	Scheduling action for IoTDs at time step t.
ϑ_t^m	RIS phase shift of the m-th element at time step t
s_t^k	State of IoTD k at time step t
$E_u(t)$	UAV u's remaining energy at time t
$b(t)$	Scheduling of SNs at time t
$j_u(t)$	Indicator for UAV u recharging at time t
$\mathbf{p}_{\text{stop}}$	Final destination of UAV
e_{th}	Energy threshold
$E_h(t)$	Energy harvested at time t (general)
ρ	Recharging price

under this category. The subsequent subsections provide detailed discussion on the states, actions, and rewards of various models in optimising multi-objectives, including trajectory, energy, scheduling, altitude, packet drop, or combinations thereof in minimising AoI. A total of twenty published works on MDPs have been reviewed under seven different categories of optimisation objectives. Meanwhile, only three published works on Partially Observable Markov Decision Processes (POMDPs) that focus on energy optimisation have been discussed, as POMDP-based problems are less common than traditional MDPs in AoI minimization for UAV-assisted IoT. At the end of each optimisation category, a summary table detailing the respective states, actions, and rewards of the MDP modeling is also provided.

2.2.1 *Trajectory Optimisation*

In this section, we discuss examples of works that have considered trajectory optimisation. Figure 2.3 presents a relevant illustration in this category based on a satellite and UAV-assisted IoT network.

2.2.1.1 Example 1

Consider a UAV-assisted IoT network where devices periodically update monitoring data for applications such as environmental sensing and industrial automation. In the absence of cellular coverage, a UAV is dispatched to fly over the network and collect

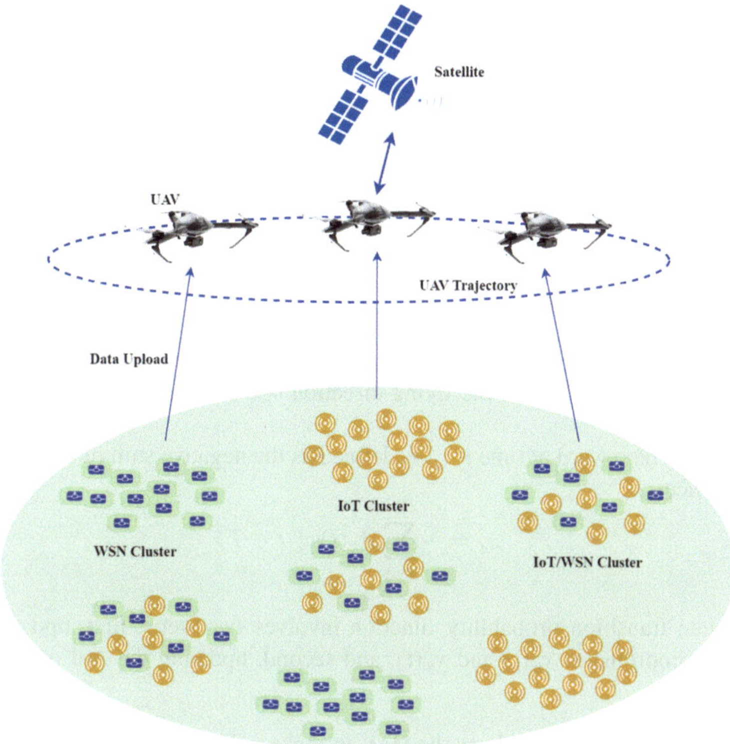

Fig. 2.3 A typical satellite and UAV-assisted IoT network model

data from the devices. The system architecture includes a satellite link to which the UAV is also connected. Given the large relative distance between the UAV and the satellite, the satellite-to-UAV link is assumed to be stable. Consequently, the time required to transmit data to the data center is considered independent of the UAV's trajectory. Moreover, due to the high probability of a line-of-sight (LoS) channel between the UAV and IoT devices, data collection is assumed to be successful as long as the UAV is within the coverage area of a device.

In this scenario, IoT traffic patterns may be either periodic or event-driven. For instance, a temperature sensor generates data at a constant sampling rate, representing a periodic pattern, whereas a fire alarm transmits data only when a temperature threshold is exceeded, illustrating an event-driven pattern. Given the presence of a large number of IoT devices, the UAV's trajectory must adapt online to accommodate these varying traffic patterns throughout its mission. To model the complex interactions between the UAV and IoT devices and to capture the dynamics of UAV movement, the trajectory planning problem is formulated as a MDP. Thus, the objective in this case is to design a novel online AoI-centered trajectory planning algorithm to minimise the AoI of IoT devices while satisfying mobility constraints based on a satellite-based UAV-IoT architecture. The problem is formulated as an MDP with the following components [8]:

- **States**: The state at time t is constituted by the horizontal ($x_u(t)$) and vertical coordinates ($y_u(t)$) of the UAV u as well as the current AoI vector $\mathbf{A}(t)$ of all IoT devices. Mathematically, the state is represented as $s_u(t) = (x_u(t), y_u(t), \mathbf{A}(t))$ for $t \in \mathcal{T}$ and $s_u(t) \in S$, where $x_u(t) \in [x_{\min}, x_{\max}]$, $y_u(t) \in [y_{\min}, y_{\max}]$, and $\mathbf{A}(t) = [A_{k_1}(t), A_{k_2}(t), \ldots, A_{k_K}(t)]$. Here, $\mathcal{T}$ denotes the set of time steps and S is the state space. Note that k denotes the index for IoT devices, and K for the total number of IoT devices.
- **Actions**: The action of UAV u at time t consists of its speed ($v_u(t)$) and flying direction ($\theta_u(t)$), denoted as $a_u(t) = (v_u(t), \theta_u(t))$. The speed $v_u(t)$ is chosen from the range $[v_{\min}, v_{\max}]$, and the flying direction $\theta_u(t)$ is chosen from the interval $(0, 2\pi]$.
- **Reward**: The reward at time t, r_t, is defined as the negative sum of the AoI of all IoT devices:

$$r_t = -\sum_{k=1}^{K} A_k(t), \quad t \in \mathcal{T}.$$

The state transition probability function involves two steps: first, updating the location coordinates $x_u(t)$ and $y_u(t)$, and second, updating the AoI of each IoT device.

This MDP formulation allows the UAV to make online decisions about its flight speed and direction based on current state information, such as the AoI of IoT devices and the UAV's location, enabling it to adapt to the dynamic network environment. The deep Q-Network (DQN)-based algorithm proposed to solve the formulated problem is provided in *Example 1* under *Value-based Solutions and Algorithms* in the next chapter.

2.2.1.2 Example 2

Consider a network setup consisting of a heterogeneous UAV swarm with varying energy capacities, and a set of distributed sensor nodes (SNs) deployed over a 2D area for environmental monitoring [6]. The UAVs are tasked with collecting data from each SN and uploading it to a data center or base station, which also serves as their initial launch point. To prevent collisions, UAVs operate at different altitudes and only initiate data acquisition by sending signals when hovering directly above a sensor. Depending on their remaining energy, a UAV may choose to either collect data or return to the data center. The flight path between two points is assumed to follow a linear trajectory, with energy consumption proportional to the travel distance.

Using multiple heterogeneous UAVs with different energy capacities to collaboratively collect data from ground sensors provides a more practical, though complex, solution. In this scenario, all UAVs depart from a common data center, with each

UAV responsible for collecting data from one or more SNs. Each SN is exclusively served by a single UAV. UAVs must determine which SNs to visit and coordinate with one another, all while ensuring sufficient energy reserves for the return trip. This framework highlights the need for efficient collaborative planning and energy-aware decision-making. The freshness of the collected data, quantified by the AoI, depends on both the time at which the data is uploaded and the time elapsed since the UAV departed from the SN.

Thus, the objective in this case is to optimise the AoI through cooperation among UAVs with different energy capabilities as they collect data from SN k. The UAV path planning process is modelled as a multi-agent MDP defined by the tuple (I, S, A, P, R, T), where I represents the set of agents (UAVs $u \in \{1, \ldots, M\}$ and sensor nodes $k \in \{1, \ldots, K\}$) that interact with the environment. Also, T represents the total mission time. Each UAV can access the states of other UAVs through information sharing. The input to the decision-making model includes the location of SN k, the location of the data center, and the current location of UAV u and the output is the UAV's flight path. The MDP is formulated as follows:

- **States**: The state for a UAV u at time t includes its position $\mathbf{p}_u(t)$ and remaining power $\eta_u(t)$. The state of sensor node k at time t, $s_k(t)$, includes its position and a binary flag indicating whether its data has been collected.
- **Actions**: The individual action of a UAV u at time t consists of a straight-line flight to the target node k when collecting data and flying back to the data center (DC). All agents start at DC with full power.
- **Reward**: An agent i receives an overall reward after the mission (at the end). The overall reward comprises two aspects: the information age reward (r^α) and the collection reward (r_c). The information age reward is related to the number of sensor nodes whose data is collected by the UAV swarm. The total reward is represented as $r = r^\alpha + r_c$, while the collection reward is given as:

$$r^c = \begin{cases} -N_c c, & c \neq 0 \\ 0, & c = 0 \end{cases} \tag{2.1}$$

When $c \neq 0$, the UAV has collected data from one or more sensors; thus, the collection reward r^c is negative. When $c = 0$, the UAV has collected data from zero sensors; thus, the collection reward r^c is zero.

This example demonstrates that with the use of DRL, UAVs can autonomously partition the mission area for collaborative data collection. In doing so, optimized and reliable cooperation among heterogeneous UAVs can be achieved to effectively minimise the AoI.

2.2.1.3 Example 3

In [9], a space-air-ground integrated network composed of IoT devices for ground-level sensing, UAVs acting as aerial data collectors, and satellites as the space-based control and communication layer was considered. UAVs fly over the IoT devices to collect data, which is subsequently transmitted to the satellite. The satellite serves as a central controller, aggregating the minimum AoI values between each UAV and IoT device, as estimated by the UAVs during training.

The primary focus of the space-air-ground integrated IoT system is on UAV trajectory design and the assignment of UAVs to IoT devices. A DRL approach is employed to construct a UAV flight model that minimises AoI and determines optimal destination locations for the UAVs. Using the collected AoI values, the satellite conducts the matching between UAVs and IoT devices, thereby defining the UAV trajectories. The UAVs operate continuously, flying to collect data and uploading it to the satellite. Thus, the objective in this case is to formulate the AoI minimization and UAV trajectory optimisation problem within SAGIN, in alignment with the reference architecture illustrated in Fig. 2.3, which includes the satellite, UAV u, and ground sensor nodes k. The components of the MDP are defined as follows:

- **States**: This is characterised by the position of UAV u at time t, i.e., $s_u(t) = (x_u(t), y_u(t))$ where $s_u(t) \in \mathcal{S}^U$.
- **Actions**: Actions of UAV u at time t include the flight distance $(d_u(t))$ and direction $(\theta_u(t))$, i.e., $a_u(t) = (d_u(t), \theta_u(t))$, $a_u(t) \in A^U$.
- **Reward**: The reward function is modeled as the negative AoI at time t for a specific IoT device $k \in \{1, \ldots, K\}$, i.e., $r_t = -A_k(t)$.

This MDP formulation addresses a layered coordination problem in which the UAV optimizes its actions based on local state information, while the satellite performs global matchmaking by pairing UAVs with IoT devices to minimise AoI. This layered approach enables the system to achieve globally minimised AoI through the combination of local decision-making and centralised coordination. The soft actor-critic algorithm proposed to solve the formulated problem is provided in *Example 1* under *Actor critic-based Solutions and Algorithms* in the next chapter.

2.2.1.4 Summary

The three published works discussed in this section focus on optimising the UAV's trajectory. Specifically, the solutions determine the optimal destination locations for the UAV(s) to minimise the AoI while collecting data from IoT devices or sensor nodes. The trajectory planning models the complex interactions between the UAV(s) and the IoT devices/sensor nodes, and is formulated as an MDP problem. A summary of the states, actions, and rewards considered in this trajectory optimisation section is presented in Table 2.2.

Table 2.2 Summary of MDP modeling for trajectory optimisation

Ref.	States	Actions	Rewards
[8]	UAV coordinates and the current AoI vector of all IoT devices	UAV speed and flying direction	Negative sum of AoI all IoT devices
[6]	UAV position and remaining power, sensor's position, and a data collection binary flag	A straight-line flight to the target node	The number of served sensors
[9]	UAV position	Flight distance and direction	Negative AoI for a specific IoT device

2.2.2 *Trajectory and Energy Optimisation*

In this section, we discuss examples of works that have considered trajectory as well as energy optimisation. Figures 2.4 and 2.5 provide relevant references to architectures for this section.

2.2.2.1 Example 1

The scenario [10] involves multiple single-antenna UAV base stations (BSs) and numerous single-antenna IoT devices, applicable to use cases such as smart grids, agriculture, safety monitoring, and geographic sensing. Each UAV operates at a

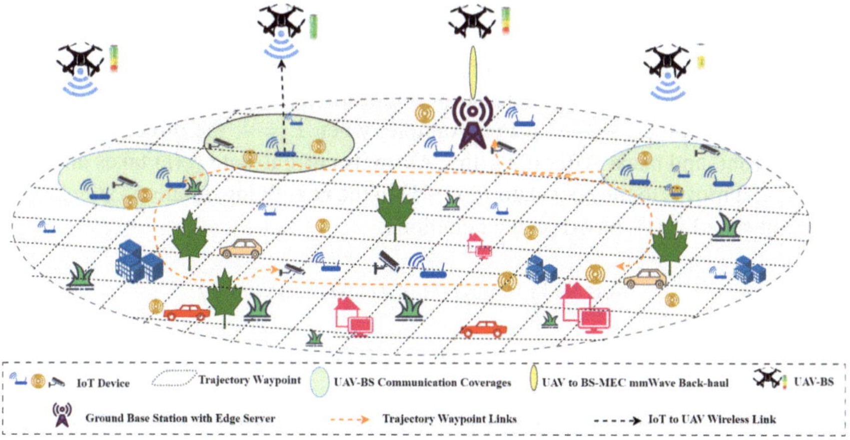

Fig. 2.4 Heterogeneous unmanned aerial networks with edge computing-assisted IoT architecture

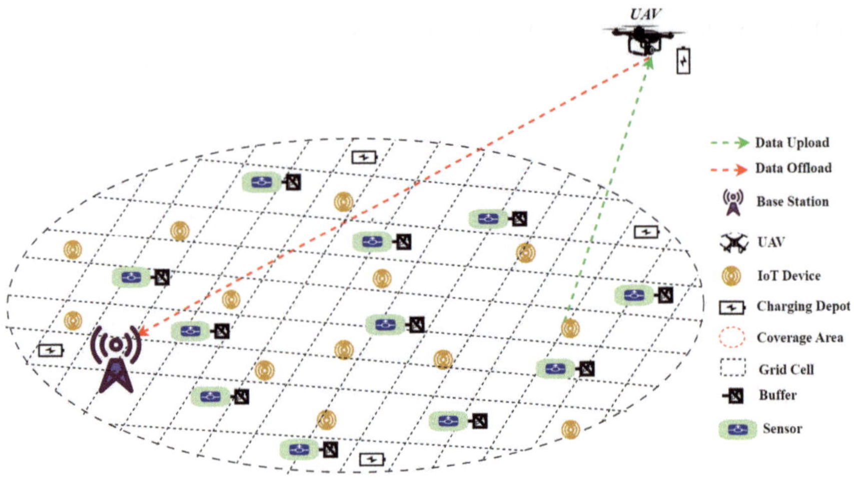

Fig. 2.5 Rotary-wing UAV-assisted IoT device network architecture

constant altitude and has a fixed sensing radius. An energy-efficiency maximisation problem is formulated under constraints related to residual energy, safety distance, and AoI. Due to the nonconvex nature of the objective function and the dynamic characteristics of the environment, the problem is modeled as an MDP rather than approached through traditional optimisation techniques. Thus, the objective in this case is to maximise energy efficiency in multi-UAV communication systems employing UAV base stations, while taking AoI constraints into account. The problem is modeled as an MDP, with each UAV acting as an agent. The MDP formulation is detailed below:

- **States**: The state of the u-th UAV at time t is described by its position, energy efficiency ($E_{eff,u}(t)$), remaining energy ($\eta_u(t)$), and the AoI of the current state ($\delta_u(t)$). Mathematically, $s_u(t) = [x_u(t), y_u(t), E_{eff,u}(t), \eta_u(t), \delta_u(t)]^T$.
- **Actions**: The action space A consists of L uniform flying directions in $(0, 2\pi]$, i.e., $A = \{0, ..., \frac{2i\pi}{L}, ..., \frac{2(L-1)\pi}{L}\}$. For example, when $L = 4$, the directions are left, right, front, and back. After the u-th UAV in its current state $s_u(t)$ takes an action $a_t \in A$, it moves to the next state $s_u(t+1)$ in the next flying duration Δt_{flight} with a flight speed $v_u(t)$, based on the equations:

$$x_u(t+1) = x_u(t) + v_u(t)\Delta t_{flight} \cos(a_t)$$

$$y_u(t+1) = y_u(t) + v_u(t)\Delta t_{flight} \sin(a_t)$$

- **Reward**: The reward function depends on the energy efficiency (see Eq. 2.2). A UAV u receives a positive reward when optimisation constraints are satisfied and obtains a penalty or no reward otherwise.

$$r_u(s_u(t), a_t) = \begin{cases} \phi\eta_u(t) & \text{if optimisation constraints are satisfied} \\ 0 & \text{otherwise} \end{cases}. \tag{2.2}$$

In this case, $0 \leq \phi \leq 1$ is a normalised energy efficiency parameter used to balance negative and positive rewards, considering that the energy efficiency value can be high.

Solving this problem using DRL is particularly promising due to the unknown and dynamic nature of the environment, which renders traditional optimisation methods ineffective. By carefully designing a reward function that reflects real-world constraints, the UAV can learn flight strategies that align with the overall network objective. DRL is suitable here because it employs neural networks to learn policies and handle the high dimensionality of the state space. Specifically, a DQN approach is adopted, incorporating experience replay and random batch sampling to stabilize learning and improve policy performance. The respective algorithm to solve the formulated problem is provided in Example 3 under *Value-based Solutions and Algorithms* in the next chapter.

2.2.2.2 Example 2

In this example [11], a trajectory design problem is considered for an uplink UAV-assisted data collection system supporting the large-scale deployment of IoT devices. The goal is to ensure timely data delivery by leveraging the UAVs' mobility, with a focus on jointly optimising power control and trajectory planning. The scenario features single-antenna UAV base stations and single-antenna IoT devices. All ground devices share subcarriers for efficient access to the wireless channel and independently transmit their data to the UAVs. Each subcarrier is assigned a fixed bandwidth, and all UAVs operate at a common flight altitude with a uniform maximum service radius. The resulting problem is a non-convex optimisation task, which is decomposed into two subtasks: power control and trajectory design. Overall, [11] aims for the joint optimisation of power control and UAV trajectory design to minimise the AoI of collected data, subject to the flying energy constraint of each UAV u. The problem is divided into two sub-tasks, both modeled using multi-agent DRL. In the power control sub-task, each subcarrier is modeled as an agent responsible for optimising its power allocation. In the trajectory design sub-task, each UAV u acts as an agent. AoI minimization is thus achieved through coordinated subcarrier power allocation and UAV trajectory planning. The MDP is formulated as follows:

- **States**: For sub-carrier power allocation, the transmit rate of each sub-carrier, $R^t_{\{k,i,l\}}$, constantly changes. The rate of the link k_i over the l-th sub-carrier at time t, $s^t_{\{k,i,l\}}$, is considered the state of the sub-carrier l, where $s^t_{\{k,i,l\}} \in S_{PC} = \{R^t_{\{k,i,l\}}\}$. For UAV u trajectory optimisation, the state space of UAV u at time t is its 3D coordinates, $s_u(t) = (x_u(t), y_u(t), z_u(t)) \in S_{TD}$.

- **Actions**: The action space for sub-carrier power allocation is the set of possible transmit powers for each sub-carrier l of link k_i, given by $P_{\{k,i,l\}} \in A_{PC} = \{\frac{P_{\max}}{L_{PC}}, 2\frac{P_{\max}}{L_{PC}}, ..., P_{\max}\}$, where L_{PC} is the length of the power control action space. The flight action of UAV u at time t is a direction chosen from $A_u(t) \in A_{TD} = \{\frac{2\pi}{L_{TD}}, 2\frac{2\pi}{L_{TD}}, ..., 2\pi\}$, where L_{TD} denotes the length of the trajectory design action space.
- **Reward**: For power control, the reward is based on the sum communication rate $R_{\text{sum},i}$ over all sub-carriers assigned to a specific link k_i. The reward function for link k_i on the l-th sub-carrier at time t is:

$$\Psi_{\text{PC},t,k,i,l} = R_{\text{sum},i} \tag{2.3}$$

For the trajectory design sub-task, the reward for UAV u is modeled as a negative function of its AoI:

$$\Psi_{\text{TD},u} = -\omega_u \lambda \tag{2.4}$$

where ω_u is the AoI value for UAV u, and λ is a weight balancing the importance of communication rate and AoI, ensuring that the AoI is minimised.

The formulated problem, involving independent power control and trajectory design, is inherently complex. To address this, a multi-agent DRL framework is proposed, with each subtask modeled as a separate MDP. This enables the deployment of a DQN solution and supports intelligent decision-making by the UAVs. The algorithm proposed to solve the formulated problem is provided in *Example 4* under *Value-based Solutions and Algorithms* in the next chapter.

2.2.2.3 Example 3

In this example [12], the navigation policy for multiple UAV-based aerial base stations (UAV-BSs) is studied, where UAVs are deployed to enhance both information freshness and connectivity for IoT devices. Battery-powered UAVs act as relays for IoT devices randomly distributed across various trajectory points. The network architecture includes a single ground base station equipped with a multi-access edge computing (MEC) server, which serves as the information fusion center. Information updates from IoT devices are relayed via UAV-BSs to the MEC server to support computation-oriented communication use cases. The MEC-equipped BS is also one of the trajectory points, and UAVs traverse these points within a finite observation time. UAV-BSs collect data from active IoT devices using conventional spectrum and directional antennas. Meanwhile, backhaul communication between UAV-BSs and the BS-MEC utilizes millimeter wave (mmWave) spectrum, supported by directional antenna arrays on both ends. The framework explicitly integrates AoI constraints and minimum energy efficiency requirements into the UAV navigation decision-making process. The solution in this case focuses on data freshness and energy-efficient optimisation of the UAV-BS u trajectory, aligned with the reference architectural framework illustrated in Fig. 2.4. The MDP formulation is as follows:

- **States**: The state space for the trajectory policy of the UAV-BS u is four-dimensional. The agent's joint observation or joint state space at each time step $t \in \{1, 2, ..., T\}$ is given by $S = \{(s_t = \mathbf{p}^u_{\text{current}}, \mathbf{p}_{\text{target}}, \eta_{BS}, \Delta_{nav}) | \eta_{BS} \in [0, \eta_{th}], \Delta_{nav} \in [1, \hat{\Delta}^{th}_b]\}$, where $\mathbf{p}^u_{\text{current}}$ represents the current location of UAV-BS u at height h_u, $\mathbf{p}_{\text{target}}$ is the target position, η_{BS} denotes the average energy efficiency (EE) of the UAV-BS, and Δ_{nav} is the average age for navigation optimisation, and $\hat{\Delta}^{th}_b$ is the AoI threshold. The trajectory position for the UAV-BS navigation u comprises the coordinates of a geographic region, i.e., $x_u \in [0, X_u]$ and $y_u \in [0, Y_u]$, where X_u and Y_u are the maximum coordinates of the geographic region. For each trial, the initial location of the UAV-BS is randomly assigned along with the number of IoTDs K. Real-world trajectory data is used to calibrate η_{BS} and Δ_{nav}.
- **Actions**: The UAV-BS $u \in \mathcal{U}$ action space involves its navigation from one allowable position to another, considering communication and trajectory constraints. The agent chooses an action a_t from the set of available actions $A = \{a_1, a_2, ..., a_U\}$.
- **Reward**: At each state transition, an immediate reward is given to the agent for forming the trajectory control policy. The immediate reward is formulated based on the instantaneous EE metric for the UAV-BS:

$$R(a_t) = \begin{cases} \alpha_1 \eta_{BS}(a_t), & \text{if trajectory, EE, and AoI constraints are satisfied} \\ -\alpha_1, & \text{if trajectory or EE constraints are violated} \\ 0, & \text{otherwise} \end{cases}$$

  The term $-\alpha_1$ penalizes the agent for violating trajectory or EE constraints. A zero reward is given when AoI constraints are violated. In this work, the agent aims to maximise the future rewards over T time slots, given as:

$$\hat{R}(s, a; t) = \sum_{t_0=0}^{T} \gamma x R_{t-t_0}(a_t),$$

  where $\gamma \in [0, 1]$ captures the trade-off between immediate and future rewards, and $R(a_t)$ changes based on the EE over the duration T as a result of the joint action (movement) from one point in the trajectory to another.

The above energy-efficient UAV-BS navigation optimisation problem formulated under contextual constraints such as energy, navigation requirements, and AoI, is an NP-hard problem. This motivates the use of DRL, specifically a Deep Q-Network with experience replay memory, to achieve energy-efficient UAV navigation under these constraints. The trained model yields an effective real-time trajectory policy for UAV-BSs that captures observable network states over time. The algorithm proposed to solve the formulated problem is provided in *Example 5* under *Value-based Solutions and Algorithms* in the next chapter.

2.2.2.4 Example 4

This example [5] considers dynamic UAV trajectory planning aimed at minimising the AoI while accounting for energy limitations in the network. IoT devices upload data using energy harvested from UAVs under dynamic channel conditions. A line-of-sight/non-line-of-sight or LoS/NLoS channel model is adopted, and a proactive UAV trajectory planning (PUTP) strategy is developed using DRL to dynamically adjust flight policies in response to changing channel states. The approach balances the trade-off between energy transfer and data collection. In this setup, each UAV is responsible for the devices within its coverage region, effectively dividing the network into distributed subsystems. Within each subsystem, a UAV wirelessly charges IoT devices and then collects their data using the harvested energy. To prevent interference, devices are not allowed to harvest energy and transmit data simultaneously. Hexagonal grids are used to model the deployment area, with each UAV's coverage region represented as a circle with a fixed radius. Hexagons are selected due to their minimal overlap properties, enabling more efficient spatial coverage and greater utilization of UAV-BSs.

To ensure fresh information delivery by UAVs to energy-constrained IoT devices, an AoI-aware UAV trajectory planning problem is formulated as a non-convex optimisation task, which is challenging to solve directly. Therefore, the problem is reformulated as a model-free MDP, accounting for uncertainties such as building diversity and weather conditions. Hexagonal grids, as shown conceptually in Fig. 2.6, are employed to minimise the overlap in network coverage, thereby maximising the utilization of UAV-BSs u. The MDP formulation defines the state, action, and reward as follows:

- **States**: The state at time slot t is represented as the index of the hexagon, $h_u(t)$, over which the UAV u is located, and the ratio $\frac{\delta(t)}{K}$ between the current AoI of all IoT devices and the planning range K , i.e., the total number of time slots considered for the UAV's scheduling and movement decisions. Mathematically, $s_u(t) = (h_u(t), \frac{\delta(t)}{K})$, where $t \in \mathcal{T}$ and $s_u(t) \in \mathcal{S}$. $\delta(t)$ represents the vector of current AoI of all IoT devices at time t, and K is the total number of IoT devices.
- **Actions**: For any UAV u at state $s_u(t)$, the action set $A(s_u(t))$ is given by $A(s_u(t)) = \text{Nei}(h_u(t)) \cup \{h_u(t)\}$, where $\text{Nei}(h_u(t))$ is the set of neighbouring hexagons of $h_u(t)$. Thus, the UAV moves to the next hexagon or remains hovering in the current hexagon. For any UAV located at state $s_u(t)$, its action set can be

$$\mathbf{a}_u(s_u(t)) = \text{Nei}(h_u(t)) \cup \{h_u(t)\}. \tag{2.5}$$

- **Reward**: The reward function within time slot t is the negative sum of AoI. This is represented in Eq. 2.6, where K is the number of IoT devices and $\delta_k(t)$ is the AoI evolution of IoT device k in time slot t.

$$r_u(t) = -\sum_{k=1}^{K} \delta_k(t). \tag{2.6}$$

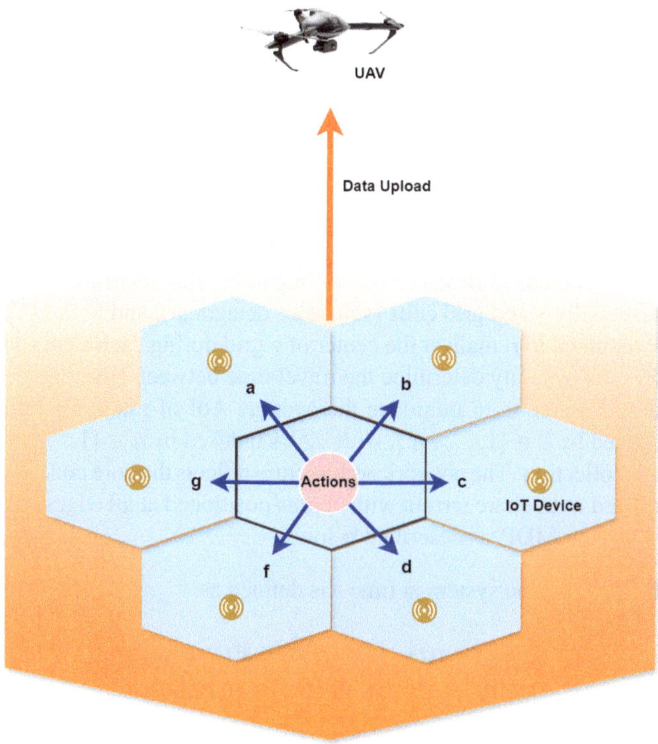

Fig. 2.6 UAV-assisted IoT spatial configuration based on a grid layout

Given the high dimensionality of the state space, neural networks are employed to extract meaningful features from the data. A DRL-based approach, specifically a DQN, is used to facilitate optimal decision-making by approximating the action-value function for each state-action pair. The algorithm proposed to solve the formulated problem is provided in Example 5 under *Value-based Solutions and Algorithms* in the next chapter.

2.2.2.5 Example 5

In this example [13], multiple UAVs are deployed to serve randomly scattered IoT devices, deploying trajectory planning with up to nine degrees of freedom. Although the scenario includes a relatively small number of sensor nodes (e.g., 10), it remains representative of practical applications such as UAV-assisted fire surveillance, where timely data collection is essential for early detection and disaster prevention. In this setup, low-power, single-antenna IoT devices are randomly distributed across a two-dimensional area to monitor various physical phenomena, with a base station located at the center. A fleet of rotary-wing UAVs is dispatched to fly over designated locations and collect data from these devices. The objective is to minimise the weighted sum of AoI while also reducing the energy consumption of each IoT device. UAV charging depots are strategically placed at fixed locations, typically at the corners of the area. A discrete-time system is considered, where UAV positions are represented by their projections onto the 2D plane. Each UAV starts its journey from a charging depot and returns there to recharge before beginning the next mission. The area is divided into equally-sized grid cells (square or octagonal), and both UAVs and IoT devices are assumed to remain at the center of a grid during each time slot. The grid spacing and UAV velocity determine the travel time between adjacent grids.

Thus, the objective is to minimise the average AoI of randomly deployed IoT devices, indexed by $k \in \{1, \ldots, K\}$, with UAVs indexed by $u \in \{1, \ldots, M\}$ responsible for data collection. The network architecture reflects the core concepts shown in Fig. 2.5, adapted to a square terrain with depots positioned at all edges. The problem is formulated as an MDP, as described below:

- **State**: The state of the system at time t is defined as

$$s(t) = (\mathbf{l}(t), \mathbf{A}(t), \boldsymbol{\delta}(t)),$$

where:

- $\mathbf{l}(t) = (\mathbf{l}_1(t), \ldots, \mathbf{l}_M(t))$ is the vector of UAV positions, with $\mathbf{l}_u(t)$ denoting the position of UAV u at time t;
- $\mathbf{A}(t) = (A_1(t), \ldots, A_K(t))$ is the vector of AoI values for the K sensor nodes;
- $\boldsymbol{\delta}(t) = (\delta_1(t), \ldots, \delta_M(t))$ is the vector of energy margins for each UAV, where $\delta_u(t)$ denotes the difference between UAV u's battery level and the energy required to reach the nearest charging depot and to relay a packet (in the worst case).

- **Actions**: The action space defines the movement direction of each UAV u at time t, represented by a direction vector $\mathbf{v}_u(t)$. Possible actions include: North (N), South (S), East (E), West (W), Northeast (NE), Northwest (NW), Southeast (SE), Southwest (SW), and Hover. These determine the UAV's next position $\mathbf{l}_u(t+1)$ as expressed in Eq. 2.7.
The full trajectory of UAV u over a trip of duration T_u is given by:

$$\mathcal{L}_u = [\mathbf{l}_u(1), \mathbf{l}_u(2), \ldots, \mathbf{l}_u(T_u)],$$

where $\mathbf{l}_u(1)$ and $\mathbf{l}_u(T_u)$ represent the initial and final positions of UAV u, respectively.

$$\mathbf{l}_u(t+1) = \begin{cases} \mathbf{l}_u(t) + (0, r) & \text{if } \mathbf{v}_u(t) = \text{North}, \\ \mathbf{l}_u(t) - (0, r) & \text{if } \mathbf{v}_u(t) = \text{South}, \\ \mathbf{l}_u(t) + (r, 0) & \text{if } \mathbf{v}_u(t) = \text{East}, \\ \mathbf{l}_u(t) - (r, 0) & \text{if } \mathbf{v}_u(t) = \text{West}, \\ \mathbf{l}_u(t) + (\frac{r}{\sqrt{2}}, \frac{r}{\sqrt{2}}) & \text{if } \mathbf{v}_u(t) = \text{Northeast}, \\ \mathbf{l}_u(t) + (-\frac{r}{\sqrt{2}}, \frac{r}{\sqrt{2}}) & \text{if } \mathbf{v}_u(t) = \text{Northwest}, \\ \mathbf{l}_u(t) + (\frac{r}{\sqrt{2}}, -\frac{r}{\sqrt{2}}) & \text{if } \mathbf{v}_u(t) = \text{Southeast}, \\ \mathbf{l}_u(t) + (-\frac{r}{\sqrt{2}}, -\frac{r}{\sqrt{2}}) & \text{if } \mathbf{v}_u(t) = \text{Southwest}, \\ \mathbf{l}_u(t) & \text{if } \mathbf{v}_u(t) = \text{Hovering}. \end{cases} \quad (2.7)$$

- **Reward**: The reward for UAV u at time t, $r_u(t)$, is defined to minimise the weighted sum of AoI of all SNs k and includes an energy penalty. Let θ_k be the weight denoting the importance of SN $k \in \{1, ..., K\}$, and let $d_{u,k}$ be the distance between UAV u and SN k. The reward is given by Eq. 2.8, where R_k is the coverage radius of SN k, and $p_{penalty}$ is the energy penalty for allocating a device outside the UAV's coverage radius.

$$r_u(t) = \begin{cases} -\sum_{k=1}^{K} \theta_k A_k(t) - p_{penalty}, & \text{if } d_{u,k} > R_k, \\ -\sum_{k=1}^{K} \theta_k A_k(t), & \text{otherwise}, \end{cases} \quad (2.8)$$

This ensures that the coverage radius R_k of each SN k is respected, and the UAV falls within its coverage radius to avoid the penalty.

The reward mechanism is designed to minimise the weighted sum of AoI by incorporating the relative importance of each sensor node, while penalizing UAVs that operate outside the effective communication range. This structure implicitly promotes both timely data collection and range-aware navigation.

2.2.2.6 Summary

This section discusses five published works focused on optimising both the UAV's trajectory and energy. The presented solutions explore diverse network architectures and constraints to achieve optimal UAV navigation alongside energy efficiency. In certain works, the UAV not only collects data from IoT devices/sensor nodes but also harvests energy, such as from a charging depot. The proposed MDP models are designed to capture the dynamic UAV movement and its intricate interactions with IoT devices/sensor nodes during these tasks, with the primary objective of minimising the AoI. Table 2.3 summarizes the states, actions, and rewards considered for trajectory and energy optimisation in this section.

Table 2.3 Summary of MDP modeling for trajectory and energy optimisation

Ref.	States	Actions	Rewards
[10]	UAV position, energy efficiency, remaining energy, AoI of the current state	Uniform flying directions	Energy efficiency over all sub-carriers assigned to a link
[11]	Sub-carrier link rate, UAV 3D coordinates	Sub-carrier transmit powers	Sum communication rate
[12]	UAV's current location, target position, energy efficiency	Navigation from one position to another	Instantaneous energy efficiency (EE) metric
[5]	UAV location, AoI ratio of all IoT devices to planning range	Set of neighboring hexagons	Negative sum of AoI
[13]	UAV positions, AoI values, energy margins	Movement direction of each UAV	Weighted sum of AoI across SNs with an energy penalty

2.2.3 *Trajectory and Scheduling Optimisation*

This section discusses examples of works that have considered trajectory and scheduling optimisation. Figure 2.7 provides a relevant abstraction to one of the architectures studied in this section.

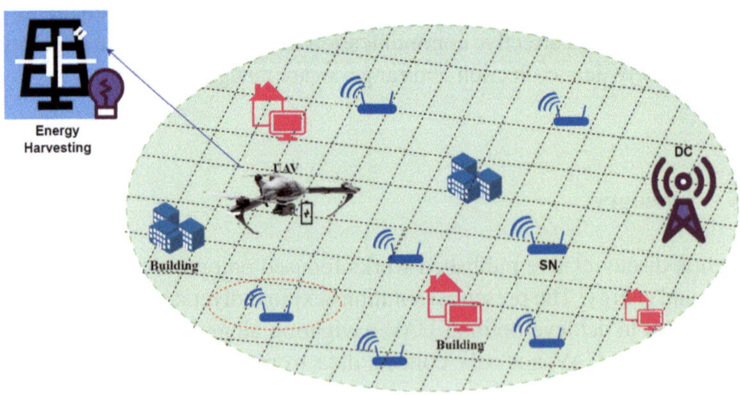

Fig. 2.7 Scene of UAV-assisted data collection, where the UAV is powered through energy harvesting

2.2.3.1 Example 1

In this scenario [14], a UAV-assisted wireless network is considered in which energy-constrained ground nodes monitor physical processes and transmit status updates. A battery-limited UAV collects these updates while navigating the region. The UAV's trajectory and the scheduling of updates are jointly optimized to minimise the weighted sum AoI across different processes. To facilitate this, the geographical region is partitioned into equal-area cells, and the UAV is assumed to fly at a fixed altitude. Its trajectory is discretized such that its position at any given time corresponds to the center of a specific cell. Thus, the aim is to design an AoI-minimal UAV-aided wireless communication system incorporating radio frequency wireless power transfer, using DRL. The UAV is modeled as the agent, while the flying environment and ground nodes constitute the environment. The problem is formulated as an MDP, and its components are described below:

- **States**: The state of the system at time step t is $s_t \in S$, denoted as $s_t = \{s_1(t), ..., s_M(t), s_u(t)\}$. The state of the m-th ground node at time step t is characterised by its battery energy and the AoI, i.e., $s_m(t) = (E_m(t), A_m(t))$. The battery capacity of each ground node is discretised into energy blocks of size E, with a maximum of $B_m^{\max}/E$ blocks. The energy for data transmission is $e_m^T(t) = \lceil \frac{E}{B_m^{\max}} E_m^T(t) \rceil$, and the harvested energy block in the t-th slot is $e_m^H(t) = \lfloor \frac{E}{B_m^{\max}} E_m^H(t) \rfloor$. The state of the UAV at time step t, $s_u(t)$, consists of its horizontal coordinate $q_u(t) = (x_u(t), y_u(t))$ and the time left $t_u(t)$ to arrive at the final location. Thus, $s_u(t) = (q_u(t), t_u(t))$. The UAV can reach the target destination within the mission time T if $t_u(t) \geq 0$; otherwise, it does not reach the target within the time period. Note that $t_u(t) = N - t - \frac{\|q_F - q_u(t)\|}{v}$, where q_F is the final location and v is the UAV's speed.
- **Actions**: The action space $A = \{a(1), ..., a(N)\}$ consists of the UAV's movement, data collection, and scheduling of ground nodes for energy harvesting (EH). The UAV chooses to move to one of the adjacent grids or hover based on five directions. The UAV's movement is $a_u(t) \in$ {North (N), South (S), East (E), West (W), Hover (H)}. The UAV's location dynamics are:

$$q_u(t+1) = \begin{cases} q_u(t) + (0, v\Delta t) & \text{if } a_u(t) = \text{N} \\ q_u(t) + (0, -v\Delta t) & \text{if } a_u(t) = \text{S} \\ q_u(t) + (v\Delta t, 0) & \text{if } a_u(t) = \text{E} \\ q_u(t) + (-v\Delta t, 0) & \text{if } a_u(t) = \text{W} \\ q_u(t) & \text{if } a_u(t) = \text{H} \end{cases}$$

  where Δt is the duration of a time slot. The action at time step t is $a_t = \{a_u(t), a_s(t)\}$, where $a_s(t)$ represents the scheduling action for ground nodes.
- **Reward/Cost**: For a policy π, the total expected cost over the finite horizon τ, starting from state s_1, is given by Eq. 2.9:

$$G^{\pi}(s_1) \triangleq \sum_{t=1}^{\tau} \sum_{m=1}^{M} \lambda_m \mathbb{E}[A_m(t)|s_1], \tag{2.9}$$

where λ_m is the importance weight of the process observed by node m, and the expectation is taken with respect to the policy π.

The problem is formulated as a finite-horizon MDP with finite state and action spaces. Due to the high dimensionality of the state space, a DRL algorithm is employed to learn the age-optimal policy that minimises the weighted sum of AoI. The next example also considers a single UAV trajectory and scheduling optimisation problem.

2.2.3.2 Example 2

This example [15] considers a UAV-assisted IoT network comprising a single BS, multiple sensor nodes, each equipped with a buffer, and a rotary-wing UAV. Unlike previous examples, this setup incorporates a queuing model at the sensor nodes, where newly sampled packets overwrite older ones. The area of interest is divided into equal-sized grids, with one sensor assumed per grid. A UAV is dispatched by the BS to fly over the sensor nodes and collect buffered data, while ensuring its residual energy remains above a critical threshold. Otherwise, it must return to the BS. The AoI at the UAV is influenced by the data sampling rate, queuing delay, and the UAV's data transmission process. To jointly optimize the UAV's trajectory and the data collection schedule under these constraints, the problem is formulated as a finite MDP and solved using a DRL approach. Overall, the objective is to optimize the UAV trajectory to ensure timely data delivery in a UAV-assisted IoT system, similar to the architecture shown in Fig. 2.5, but involving a single UAV. The system is modeled as a finite MDP, and DRL is employed to derive the optimal flight path while accounting for environmental factors. The components of the MDP are defined as follows:

- **States**: The state space at time step t is described as $s_t = (A_t, U(t), q_u(t), e(t))$, where:

  1. $A_t = (A_1(t), A_2(t), ..., A_K(t)) \in \prod_{k=1}^{K} \mathcal{A}_k$ is the AoI vector of K sensors.
  2. $U(t) = (U_1(t), U_2(t), ..., U_K(t)) \in \prod_{k=1}^{K} \mathcal{U}_k$ is the vector of lifetimes of update packets in the SN buffers, with $\mathcal{U}_k = \{0\} \cup \mathcal{A}_k$.
  3. $q_u(t) = (x(t), y(t))$ is the horizontal location of the UAV.
  4. $e(t) \in [0, E_{\max}]$ is the residual energy of the UAV.

- **Actions**: The UAV movement is limited to one of its four adjacent grid cells: West (W), East (E), North (N), and South (S). Thus, the action space is $A = \{\text{W, E, N, S}\}$. The UAV position changes according to Eq. 2.10:

$$q_u(t+1) = \begin{cases} q_u(t) + (0, L) & \text{if } a_t = \text{N} \\ q_u(t) + (-L, 0) & \text{if } a_t = \text{W} \\ q_u(t) + (L, 0) & \text{if } a_t = \text{E} \\ q_u(t) + (0, -L) & \text{if } a_t = \text{S} \end{cases} \tag{2.10}$$

where L is the length of a grid cell.

- **Reward**: The cost function is modeled as the weighted sum of the sensors' average AoI, packet loss rate, and energy consumption, as shown in Eq. 2.11:

$$r_t = \frac{1}{K} \sum_{k=1}^{K} (A_k(t) + \delta_1 \cdot m_k(t)) + \delta_2 \cdot e_c(t) \tag{2.11}$$

where $m_k(t)$ is the packet loss rate of sensor k at time t, $e_c(t)$ is the energy consumption at time t, δ_1 is the weight for the packet loss rate, and δ_2 is the weight for energy consumption.

To address the challenge of high-dimensional state and action spaces, a Double Deep Q-Network (DDQN) algorithm is proposed to derive an age-optimal UAV trajectory while simultaneously minimising the packet drop rate. The algorithm proposed to solve the formulated problem is provided in Example 6 under *Value-based Solutions and Algorithms* in the next chapter.

2.2.3.3 Example 3

This example [7] considers a UAV-assisted wireless network in which a battery-constrained UAV collects status updates from energy-limited ground nodes monitoring diverse physical processes (e.g., agricultural, healthcare, or industrial systems). Given the constrained energy budgets of the ground nodes, only small data packets are transmitted, and each device operates with limited battery capacity. The UAV maintains a fixed flight altitude, and time is discretized into slots. The optimisation problem is initially formulated as a mixed-integer program. To tackle the combinatorial complexity of joint scheduling and trajectory planning, a reinforcement learning framework is adopted. Specifically, the problem is modeled as a finite-horizon MDP and solved using a neural combinatorial deep reinforcement learning (NCRL) algorithm. To ensure scalability in large networks, a Long Short-Term Memory (LSTM)-based autoencoder is introduced to map high-dimensional state representations into fixed-size vectors while capturing the spatio-temporal dependencies between update locations and their associated time instants. Thus, the aim is to optimize the UAV trajectory and status update scheduling of ground nodes for normalized weighted sum of AoI (NWAoI) minimization. The UAVs aim to maintain high information freshness across the geographical region, supporting applications such as agriculture, public safety, healthcare, and industrial monitoring. The authors formulate the problem as an MDP as follows:

- **States**: The state S_t is represented as a matrix with $K+1$ rows and $t+1$ columns. The first column contains the initial battery levels of the K nodes and the initial time instant (set to 0). Each subsequent column $i > 1$ corresponds to the system state after the i-th update: it includes the energy levels of all nodes and the time t_i at which the update occurs. Formally, the i-th column of S_t is defined as $\mathbf{s}_{t,i} \triangleq [E_1(t_i), ..., E_K(t_i), t_i]^T$. The initial state is $S_0 = \mathbf{s}_{0,1} = [E_{\max,1}, ..., E_{\max,K}, 0]^T$. The state space consists of all such matrices where each element in row $k \in \{1, ..., K\}$ is within $[0, E_{\max,k}]$, and the last row (time) is within $[0, \tau]$, where τ is the total operational time.
- **Actions**: At any state S_{t-1}, the UAV can either schedule a node $k \in \{1, ..., K\}$ to send an update or terminate the current scheduling policy. Formally, the action a_t is an integer in the action set $\mathcal{A} \triangleq \{0, 1, ..., K\}$, where $a_t = 0$ indicates termination, and $a_t = k > 0$ schedules node k to transmit. Each action appends a node to the current scheduling policy vector, which updates the system state.
- **Reward**: When an action is taken, the system transitions to a new state, and the UAV receives a reward based on the decrease in normalized weighted AoI (NWAoI). The reward associated with action a_t at state S_{t-1}, given the previous policy u_{t-1}, is:

$$r_t(S_{t-1}, u_{t-1}, a_t) = \bar{G}(u_{t-1}) - \bar{G}(u_t),$$

  where $u_t = [u_{t-1}, a_t]$ is the updated scheduling policy. We define $\bar{G}(u_0) = 1$, representing the worst-case NWAoI when no nodes are scheduled. The reward for the termination action ($a_t = 0$) is defined as zero.

The MDP is designed such that, at each state, the UAV appends a new node to the end of the scheduling policy, solves the corresponding NWAoI minimization problem, and transitions to a new state. During this transition, the UAV obtains an updated NWAoI value, which is used as the reward signal to guide the learning process toward the optimal scheduling policy. The DQN algorithm proposed to solve the formulated problem is provided in *Example 7* under *Value-based Solutions and Algorithms* in the next chapter.

2.2.3.4 Example 4

In this example [16], a UAV operating in half-duplex mode is deployed as a mobile data collector, gathering data from IoT devices and subsequently offloading it to the base station. These IoT devices must continuously upload sampled data to the network. However, due to limited energy resources and poor channel conditions, they are unable to communicate directly with the remote base station, thus necessitating the deployment of UAVs. In half-duplex mode, each device can either transmit or receive data in a given time slot, but not both. As a result, data collection and forwarding cannot occur simultaneously within a coherent interval. This work investigates an AoI- and energy-aware data collection framework for UAV-assisted IoT networks, aiming to minimise the weighted sum of the expected average AoI, UAV propulsion energy, and the transmission energy of IoT devices. To achieve this, the UAV's flight speed, hovering locations, and bandwidth allocation for data collection are jointly optimized. Given the dynamic nature of the system, the problem is modeled as an MDP. Thus, the expected AoI and energy consumption minimization problem for energy-aware UAV-aided data collection in IoT is formulated as a discrete-time MDP with an infinite-horizon discounted criterion, in which the UAV is regarded as the agent. The main components of the MDP are as follows:

- **States**: The state at time step t is characterised by the UAV's location, the current AoI of all IoTDs, $A_t = (A_1(t), ..., A_K(t))$, and the coverage condition indicator $C_m^t \in \{0, 1\}$, which indicates whether IoTD m has been served in the t-th Hovering-Traveling Process (HTP) of the current flight round. Time is slotted in T, and one HTP includes a hovering duration and a flying duration.
- **Actions**: The action space includes the flight distance (within trajectory bounds), the UAV flight direction $\theta_t \in (0, 2\pi]$, and the UAV speed v_t.
- **Reward**: During the t-th HTP, the reward r_t is related to the average AoI ($\bar{A}_t$), IoTD energy consumption (E_{IoTD}^t), and UAV propulsion energy (E_{UAV}^t), as given by Eq. 2.12:

$$r_t = -\bar{A}_t - \zeta \cdot E_{\text{IoTD}}^t - \omega \cdot E_{\text{UAV}}^t - h_{\text{punish}} + h_{\text{award}} \tag{2.12}$$

  where ζ and ω are weight factors for the IoTDs' transmission energy and the UAV's propulsion energy, respectively. A penalty $h_{\text{punish}} > 0$ is incurred if the UAV cannot serve scheduled IoTDs in a time round, and a reward $h_{\text{award}} > 0$ is given when the UAV covers all IoTDs in one sub-area.

To address the multidimensional action space inherent in UAV trajectory planning, a Twin Delayed Deep Deterministic(TD3)-based algorithm is proposed. TD3 algorithm is a policy gradient method inspired by the Deep Deterministic Policy Gradient (DDPG). A deep neural network is employed for feature extraction to support efficient policy learning. By tuning the energy weight factors, the scheme enables a trade-off among the expected AoI, the transmission energy of IoT devices, and the UAV's propulsion energy, achieving energy-aware and AoI-efficient trajectory optimisation. The algorithm proposed to solve the formulated problem is provided in Example 2 under *Actor-Critic-based Solutions and Algorithms* in the next chapter.

2.2.3.5 Example 5

This example [17] explores a UAV-assisted wireless sensor network in which, for the first time among the examples considered, the UAV is powered by energy harvesting to support sustainable operation. The UAV acts as a mobile relay, collecting data from randomly deployed energy-constrained ground sensor nodes and forwarding it to a fixed-location data center. A rotary-wing UAV with a limited-capacity battery is deployed, and the coverage area is divided into a grid of uniform cells. The UAV operates in two modes: a working mode, where it collects and forwards data, and a waiting mode, where it lands to harvest energy when its residual battery level drops below a certain threshold. Once sufficient energy has been harvested, it resumes its operation. Both the UAV's trajectory and the transmission scheduling of sensor nodes are jointly optimized to minimise the average AoI while reducing the UAV's energy consumption. Overall, the aim is to address UAV-assisted data collection in wireless sensor networks using energy harvesting for sustainable UAV operation, as conceptually illustrated in Fig. 2.7. The approach jointly optimizes the transmission opportunities of ground SNs and the UAV's flight trajectory to minimise AoI while reducing energy consumption. The MDP is formulated as follows:

- **States**: For the UAV at time step t, the state is $s_u(t) \triangleq [L_u(t), E_u(t)]$, where $L_u(t) = (x_u(t), y_u(t))$ is its position and $E_u(t)$ is its residual energy. For sensor node k at time step t, the state is $s_k(t) \triangleq [A_k(t), U_k(t), q_k(t)]$, where $A_k(t)$ is the AoI, $U_k(t)$ is the lifetime of the latest update packet, and $q_k(t)$ is the queue length.
- **Actions**: The UAV takes different actions in working and waiting modes. At time step t, the action is $a_t = [a_f(t), \pi(t)]$, where $a_f(t) = [a_h(t), a_v(t)] \in \{-L, ..., L\} \times \{-L, ..., L\}$ represents the horizontal $(a_h(t))$ and vertical $(a_v(t))$ flight displacements in terms of grid cells, with a maximum displacement of L due to velocity constraints. $\pi(t) \in \{0, 1, ..., K\}$ indicates the number of grids the UAV flies above. In the working mode, the UAV selects horizontal flight displacements, and data uploading occurs at the start of each time slot. The action space in this mode is $\mathcal{A} = \{-L, ..., L\} \times \{-L, ..., L\} \times \{0, 1, ..., K\}$. In the waiting mode (when energy is below a threshold), the UAV performs vertical flight actions (ascending, descending, standing still) and has no horizontal movement, waiting for harvested energy. It switches back to working mode when sufficient energy is harvested. The UAV's location is updated based on $L_u(t + 1) = L_u(t) + [a_h(t)x_s, a_v(t)y_s]$, where x_s and y_s are the grid dimensions, and $v_h(t)$ and $v_v(t)$ are the horizontal and vertical speeds.
- **Reward**: The cost function is the weighted sum of the average AoI of nodes and the UAV's energy consumption. The objective is to find the optimal policy to minimise the cumulative cost. The cost function at time step t is $c(t) = \overline{A(t)} + \zeta e(t)$, where $\overline{A(t)}$ is the mean AoI of all SNs, ζ is a weight coefficient, and $e(t)$ is the UAV's energy consumption in that slot. The cost depends on the state s_t and action a_t, i.e., $C(s_t, a_t)$.

Given the high-dimensional nature of the joint state' action space and the absence of known state transition probabilities, traditional MDP solution methods such as value iteration and policy iteration are unsuitable for this problem. To address these limitations, the Asynchronous Advantage Actor-Critic (A3C) algorithm is employed within a DRL framework. A3C enables efficient policy learning and faster convergence by leveraging asynchronous updates across parallel agents, making it well-suited for the complex control environment of energy-harvesting UAV-assisted data collection. The algorithm proposed to solve the formulated problem is provided in Example 3 under *Actor critic-based Solutions and Algorithms.*

2.2.3.6 Summary

This section examines five published works that aim to optimize both the UAV's trajectory and scheduling. Different scenarios with varying constraints and parameters are considered to determine the optimal UAV navigation and scheduling for data collection tasks. For instance, one work (Example 4) considered the trade-off among expected AoI, IoT device transmission energy, and the UAV's propulsion energy in its scheduling optimisation. In this case, the proposed MDP model aims to capture the system's dynamics to optimize the UAV's flight speed, hovering locations, and bandwidth allocation for data collection. A summary of the states, actions, and rewards considered in this trajectory and scheduling optimisation section is provided in Table 2.4.

2.2.4 Trajectory, Energy, and Scheduling Optimisation

In this section, we discuss examples of works that have considered trajectory, energy, and scheduling optimisation. Figure 2.8 presents a typical UAV-assisted data collection architecture referenced in this section.

Table 2.4 Summary of MDP modeling for trajectory and scheduling optimisation

Ref.	States	Actions	Rewards
[14]	Ground node battery energy, AoI of each node	UAV's movement, data collection, scheduling of ground nodes	Total expected cost over a finite horizon
[15]	AoI vector, packet lifetime vector, UAV's horizontal location, UAV residual energy	UAV movement in four directions	Weighted sum of average AoI, packet loss rate, energy consumption
[7]	Initial battery levels, updated energy levels after each time instant	Terminate current schedule or schedule a node to send an update packet	Maximum NWAoI from chosen action
[16]	UAV location, current AoI of all IoTDs, coverage condition	Flight distance, UAV flight direction, UAV speed	Average AoI, IoTD energy consumption, UAV propulsion energy
[17]	UAV position, UAV residual energy, sensor AoI, update packet lifetime, queue length	Flight displacements	Weighted sum of: average AoI of nodes and UAV energy consumption

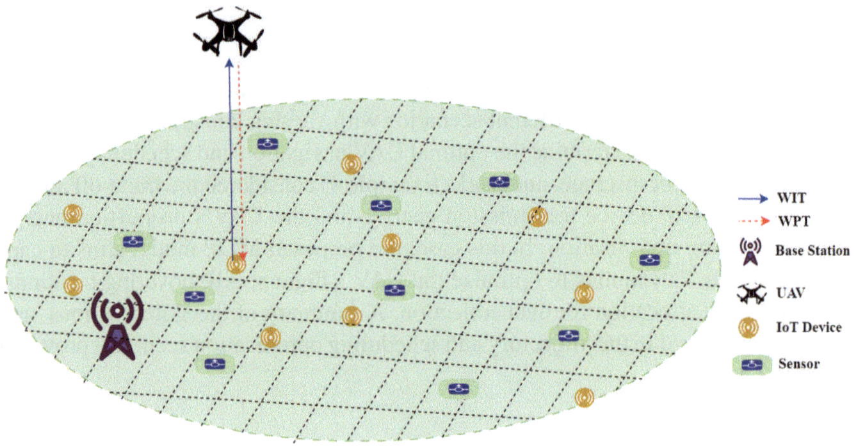

Fig. 2.8 Typical architecture for UAV-assisted data collection

2.2.4.1 Example 1

This example [18] explores a UAV-assisted wireless-powered communication network in which a UAV is dispatched to wirelessly charge multiple energy-constrained ground nodes (GNs) using radio frequency (RF) energy transfer. These GNs, deployed to sense environmental parameters such as illumination, temperature, and humidity, harvest and store energy for later use in uplink data transmission to the UAV. In each time slot, the UAV either charges the GNs or schedules one GN to upload its sensed data. The UAV operates at a constant altitude within a discretized flying region partitioned into small grids, with its movement constrained to one grid per time slot to ensure smooth trajectory transitions. The communication protocol follows a TDMA scheme to avoid interference between energy transfer and data transmission, with GNs allowed to transmit only when sufficient energy has been accumulated.

The core objective is to minimise the system's average AoI by jointly optimising the UAV's flight trajectory, energy harvesting schedule, and data transmission timing. Due to the nonconvex nature of the resulting optimisation problem, exacerbated by binary scheduling decisions and coupled trajectory-energy variables, it becomes computationally intractable as the number of GNs increases. However, since the system exhibits the Markov property (i.e., the next state depends only on the current state), the problem is modeled as a finite-horizon MDP with defined state and action spaces. This work is among the first to tackle UAV-assisted wireless-powered data collection for status update applications using DRL.

The underlying network architecture is conceptually illustrated in Fig. 2.8. The model incorporates Wireless Power Transfer (WPT) from UAVs to SNs and Wireless Information Transfer (WIT) from SNs to UAVs. As SN density increases, solving the non-convex optimisation problem becomes computationally prohibitive, making traditional methods impractical. Since trajectory, AoI, and energy are independent, and the next state depends solely on the current state, the problem can be modeled as a Markov process with finite state-action spaces as follows:

- **States**: The state comprises the states of the ground nodes and the UAV. The state of the k-th SN at time step t depends on the energy stored in its battery and its AoI: $s_k(t) = (E_k(t), A_k(t))$. The battery capacity is divided into energy blocks. The state of the UAV, $s_u(t)$, consists of its horizontal coordinates $q_u(t) = (x_u(t), y_u(t))$ and the remaining time $t_u(t)$ to reach the final location. Thus, $s_u(t) = (q_u(t), t_u(t))$.
- **Actions**: The UAV action is characterised by its movement, data collection, and ground node energy harvesting scheduling. In each time slot, the UAV either moves to a new grid or hovers above its current grid. For movement, the UAV can move North (N), South (S), East (E), West (W), or Hover (H). Mathematically, $a_u(t) \in \{\text{N, S, E, W, H}\}$. The UAV's location dynamics are described by Eq. 2.13:

$$q_u(t+1) = \begin{cases} q_u(t) + (0, v\Delta t) & \text{if } a_u(t) = \text{N} \\ q_u(t) + (0, -v\Delta t) & \text{if } a_u(t) = \text{S} \\ q_u(t) + (v\Delta t, 0) & \text{if } a_u(t) = \text{E} \\ q_u(t) + (-v\Delta t, 0) & \text{if } a_u(t) = \text{W} \\ q_u(t) & \text{if } a_u(t) = \text{H} \end{cases} \tag{2.13}$$

  where v is the UAV speed and Δt is the duration of a time slot.
- **Reward**: The reward function is designed to minimise the average AoI by jointly adjusting the UAV's trajectory and the transmission scheduling of the ground nodes. The reward function considers the AoI of the packets collected by the UAV and the time left to reach the final location, $t_u(t)$. The UAV receives an extra reward if it has sufficient time to reach the final location and is penalised otherwise, as shown in Eq. 2.14:

$$r(t) = \begin{cases} (k_1 z(t) + k_2 t_u(t))t & \text{if } t_u(t) < 0 \\ (k_1 z(t) + k_2 t_u(t) + k_3)t & \text{otherwise} \end{cases} \tag{2.14}$$

  where $z(t) = -\sum_{k=1}^{K} A_k(t)$, and k_1, k_2, and k_3 are positive constants. Here, k_1, k_2, and k_3 are weights associated with the AoI, time-to-goal, and a late-arrival penalty, respectively.

To solve the formulated optimisation problem, a Deep Q-Network (DQN) framework is adopted to approximate the optimal policy. Leveraging the representational power of neural networks, the proposed scheme effectively handles the large state space and discrete action set. Simulation results demonstrate that the DQN-based solution significantly outperforms baseline methods, achieving a much lower average AoI.

2.2.4.2 Example 2

This example investigates a rechargeable UAV-assisted IoT network, where the UAV acts as an edge server for delay-sensitive monitoring tasks by performing both data collection and processing. Due to its mobility and cost-effectiveness, the UAV is well-suited for such applications; however, its limited battery capacity poses a key operational challenge. To address this, the scenario assumes that the UAV can recharge its energy via wireless power transfer from fixed CSs distributed throughout the network. The UAV departs from a designated initial location, collects data packets from ground SNs, and must reach its final destination within a predefined mission time. The operational area is divided into a grid of small, equally sized cells, and the UAV is constrained to move between adjacent grids in each time slot to ensure smooth trajectory planning. Each SN, equipped with a single antenna, transmits environmental data (e.g., temperature, humidity, illumination) to the UAV when it is within communication range and has sufficient stored energy. To jointly minimise the weighted sum of average AoI and average energy recharging cost, a finite-horizon MDP is formulated, which governs the UAV's trajectory, SN scheduling, and energy replenishment decisions. Thus, this example considers a UAV-assisted WSN architecture with CSs, based on an adapted version of the conceptual architecture in Fig. 2.5. An energy-efficient AoI-aware UAV-based data collection problem [19] is reformulated as an MDP. A DRL-based algorithm is used to minimise the weighted sum of average AoI and average recharging cost. The MDP is formulated as follows:

- **States**: At time step t, the state s_t depends on the UAV's projection on the ground $\mathbf{o}(t)$, the vector of AoIs $\mathbf{A}(t)$ for all SNs, the time difference $\phi(t)$ between the remaining and the minimum required time to reach the final location, and the UAV's remaining energy $E_u(t)$. Thus, $s_t = (\mathbf{o}(t), \mathbf{A}(t), \phi(t), E_u(t))$.
- **Actions**: The UAV's action a_t at time step t depends on its movement $v(t) \in$ {North, South, East, West, Hovering}, the scheduling of SNs $b(t) \in \{0, 1, ..., K\}$, and a recharging indicator $j_u(t) \in \{0, 1\}$. Hence, $a_t = (v(t), b(t), j_u(t))$.
- **Reward**: The reward function aims to minimise the weighted sum of the average AoI and the average recharging price. The UAV is rewarded if it reaches the final destination at the expected time T with sufficient remaining energy (i.e., $E_u(T) \geq e_{th}$), and penalized otherwise. The reward function is given by:

$$r_t = \begin{cases} -G(t) - r_1 & \text{if } \phi(t) < 0 \text{ or } E_u(t) < e_{th} \\ -G(t) + r_2 & \text{if } \mathbf{o}(T) = \mathbf{p}_{\text{stop}} \text{ and } E_u(T) \geq e_{th} \\ -G(t) & \text{otherwise} \end{cases}$$

where $G(t) = \frac{1}{T}(\omega_1 \sum_{k=1}^{K} A_k(t) + \omega_2 \rho E_h(t))$, with ω_1 and ω_2 as the weight coefficients for the AoI and recharging price, respectively, and $E_h(t)$ denoting the energy harvested at time t.

To address the formulated optimisation problem, a dueling double deep Q-network (D3QN) is trained to jointly optimize the UAV's flight trajectory, sensor node scheduling, and energy recharging strategy. This learning-based approach effectively captures the trade-offs between data freshness, energy availability, and mission deadlines. Simulation results demonstrate that the proposed D3QN-based algorithm achieves superior performance compared to conventional distance-based and AoI-based baselines.

2.2.4.3 Summary

This section discusses two works focused on optimising UAV trajectory, energy, and scheduling. The solutions presented consider different network architectures with varying constraints to determine the optimal scheduling for UAV task completion, while accounting for energy efficiency requirements. One of these works aims to minimise the system's average AoI by jointly optimising the UAV's flight trajectory, energy harvesting schedule, and ground node data transmission timing. The proposed MDP in this case captures the dynamics of UAV movement and its interactions with the nodes in performing both data collection and energy harvesting tasks. A summary of the states, actions, and rewards considered in this trajectory, energy, and scheduling optimisation section is provided in Table 2.5.

2.2.5 Trajectory, Scheduling, and RIS Phase Shift Optimisation

This section discusses an example that considers trajectory, scheduling, and RIS phase shift optimisation. Figure 2.9 presents a reference architecture relevant to this section.

Deploying UAVs to support IoT data collection is particularly critical in urban scenarios where real-time information is essential. In this example [27], an urban environment is considered, where a rotary-wing UAV collects status updates from spatially distributed IoT devices. Due to building-induced blockages, traditional line-

Table 2.5 Summary of MDP modeling for trajectory, energy, and scheduling optimisation

Ref.	States	Actions	Rewards
[18]	Sensor's energy, sensor's AoI, UAV's horizontal coordinates, remaining time to destination	UAV movement, data collection, energy harvesting scheduling	Average AoI of packets collected by the UAV
[19]	UAV's projection, vector of AoIs for all SNs, UAV's remaining energy	UAV movement, scheduling of SNs, recharging indicator	Weighted sum of: average AoI and average recharging price

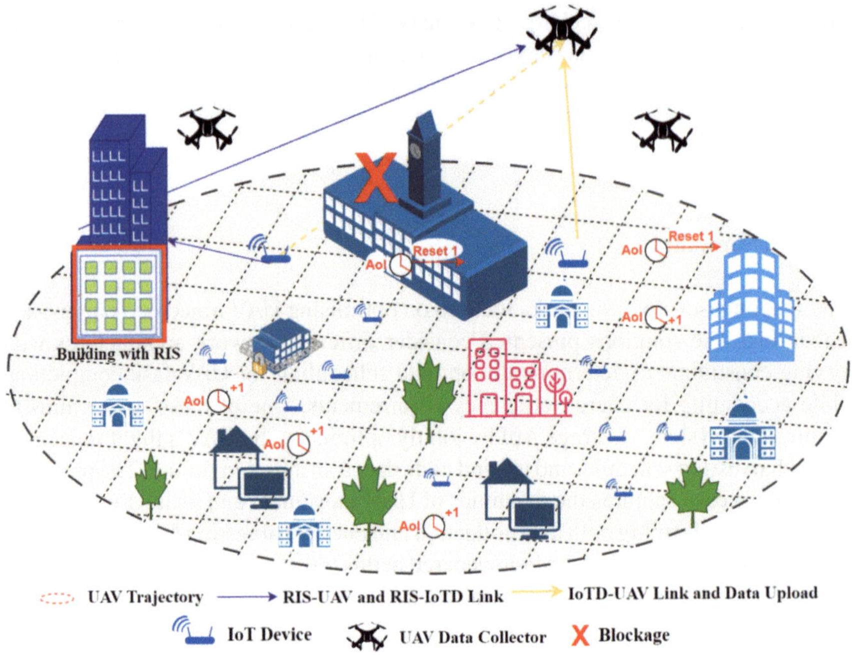

Fig. 2.9 RIS-assisted UAV data collection scenario

of-sight communication is often unreliable. To address this, a reconfigurable intelligent surfaces (RIS) equipped with a uniform linear array of reflecting elements is installed on the surface of high-rise buildings to assist signal propagation. The system is discretized into equal-length time slots, and a TDMA protocol is employed to avoid inter-user interference, such that only one IoT device is allowed to transmit an update in each time slot. Unlike conventional approaches relying on probabilistic LoS models, this work adopts a more realistic urban channel characterization, where link connectivity between the UAV and IoT devices is determined by the physical environment.

Overall, the aim is to minimise the average AoI of all IoT devices by optimising the UAV flight trajectory, IoT device transmission scheduling, and RIS phase shift configuration, based on the system architecture conceptually illustrated in Fig. 2.9. The problem is formulated as an MDP. The definitions of state, action, and reward of the MDP are provided below:

- **State**: The state at time step t is given by the UAV's coordinates $q_u(t)$ and the AoI of all IoTDs A_t. Mathematically, $s_t = [q_u(t), A_t]$, where $A_t = (A_t^1, A_t^2, ..., A_t^K)$.
- **Action**: The action at time step t includes the UAV's average flying speed v_t, the horizontal direction of the UAV θ_t, and the transmission scheduling policy of the IoTDs $\{k \in K | s_t^k = 1\}$. Once the UAV takes an action, the RIS controller manipulates the RIS elements to adjust their phase shift ϑ_t^m, thus improving com-

munication quality between the UAV and the scheduled IoTD while respecting all relevant optimisation constraints.

- **Reward**: To maximise the reduction in AoI of IoTDs, the reward captures and measures the effect of the AoI update process. The reward at time step t is given by Eq. 2.15:

$$r_t = \frac{1}{K} \sum_{k=1}^{K} (A_t^k - A_{t+1}^k) + \delta_p, \tag{2.15}$$

where δ_p is a penalty incurred whenever the UAV takes an incorrect move, e.g., leaves the target area.

The formulated problem involves hybrid decision variables comprising both high-dimensional continuous and discrete components, which makes traditional optimisation intractable. To address this, a hybrid DRL-based method is proposed. Specifically, a Soft Actor-Critic (SAC) algorithm is employed due to its stability and scalability in continuous action spaces, and its superior performance over traditional approaches like DDPG. The SAC framework integrates off-policy training with a stochastic actor and incorporates prioritized experience replay to accelerate convergence and improve learning stability. In parallel, an alternating optimisation (AO) method is applied to configure the phase shifts of the RIS. The resulting SAC-AO-RIS framework jointly optimizes the UAV's trajectory and transmission scheduling, while dynamically tuning the RIS to enhance link quality in the urban environment. This hybrid approach effectively manages the challenges posed by mixed-variable, non-convex, and partially observable urban settings while maintaining low AoI. The algorithm proposed to solve the formulated problem is provided in Example 4 under *Actor Critic-based Solutions and Algorithms* in the next chapter.

2.2.6 Scheduling Optimisation and Altitude Control

In this section, we discuss examples of works that have considered scheduling optimisation and altitude control. Figures 2.10 and 2.11 provide typical reference architectures relevant to the examples in this section.

2.2.6.1 Example 1

This example introduces the integration of RIS with UAVs to enable energy-efficient and information-fresh wireless communication. RIS technology enhances the propagation environment by intelligently adjusting the phase shifts of its reflective elements to boost desired signal strength and suppress interference. When combined with UAVs, RIS can improve signal quality and extend communication coverage in dynamic IoT environments. In the considered scenario, a single UAV equipped with RIS is deployed as a passive relay to forward data sampled by IoT devices (IoTDs) to

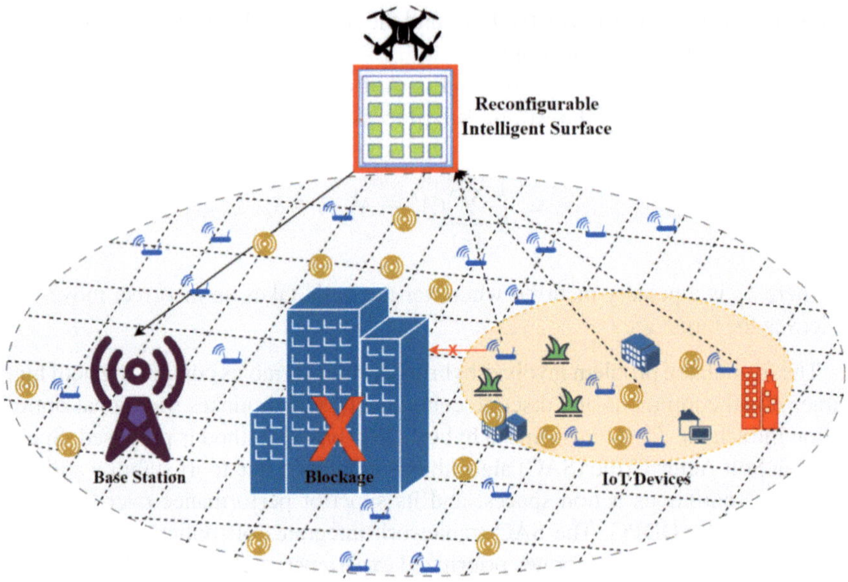

Fig. 2.10 Aerial RIS-assisted IoT network architecture

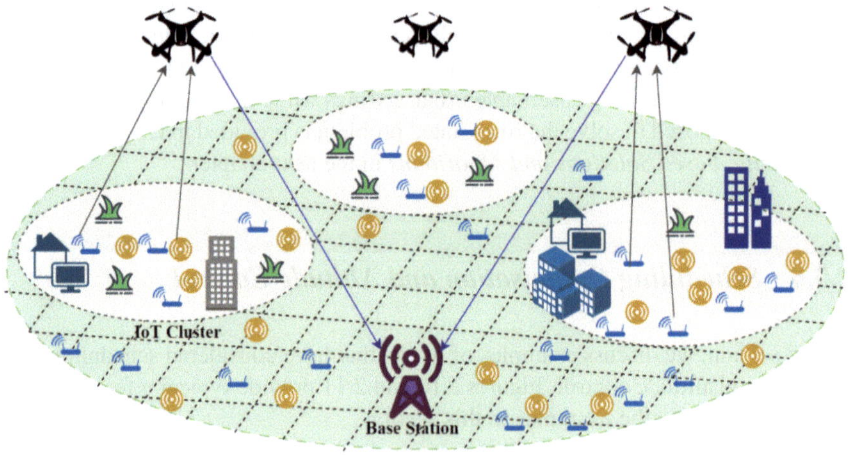

Fig. 2.11 UAV-assisted IoT network with clustered users

a BS. These devices monitor stochastic processes and transmit their data, which must be received at the BS with high freshness. Data transmission is considered successful only if the signal-to-noise ratio (SNR) exceeds a given threshold, upon which the AoI decreases.

The objective is to jointly optimize the UAV's altitude, communication scheduling, and RIS phase shifts to minimise the expected sum AoI (ESA), subject to constraints including UAV altitude bounds, SNR thresholds, and IoTD scheduling feasibility.

While the locations of IoTDs and the BS are assumed to be known in advance, the activation patterns of the IoTDs are unknown and vary over time, reflecting practical constraints in many real-world applications. The region is discretized into uniform grids, and the UAV can adjust its altitude dynamically at each time slot within a specified range. The BS acts as a centralised controller that manages UAV altitude, RIS configuration, and IoTD scheduling via reliable control links. Given the stochastic and high-dimensional nature of the problem, which includes both discrete and continuous variables, the resulting mixed-integer non-convex optimisation problem becomes intractable with conventional methods.

Thus, in this example [20], the aim is to formulate the AoI minimization problem for a RIS-carried UAV as an MDP, focusing on UAV altitude adaptation and scheduling policy control. Figure 2.10 provides a reference conceptual architecture relevant to the RIS-carried UAV scenario. To optimize the RIS configuration and minimise AoI, the problem is modeled as an MDP, formulated as follows:

- **States**: At time step t, the system state is $s_t = (\mathbf{A}(t), \Upsilon(t), z_u(t))$, where $\mathbf{A}(t) = (A_1(t), \ldots, A_K(t))$ is the vector of AoI values for the K IoT devices, $\Upsilon(t) = (\Upsilon_1(t), \ldots, \Upsilon_K(t))$ is the SNR vector resulting from RIS-assisted signal combination, and $z_u(t)$ denotes the UAV's altitude.
- **Actions**: The RL agent executes an action $a_t = (\xi(t), \Delta h(t))$ at each time step. The vector $\xi(t) \in \{0, 1\}^K$ is a binary scheduling indicator where $\xi_k(t) = 1$ if IoTD k is scheduled to transmit, and 0 otherwise. The UAV altitude adjustment is represented by $\Delta h(t) \in \mathcal{H} = \{(1, 0, 0), (0, 1, 0), (0, 0, 1)\}$, where:

  - $(1, 0, 0)$: UAV increases its altitude,
  - $(0, 1, 0)$: UAV decreases its altitude,
  - $(0, 0, 1)$: UAV hovers at the same altitude.

- **Reward**: The immediate reward at time step t is defined as:

$$r_t = -\sum_{k=1}^{K} A_k(t)$$

  encouraging the agent to minimise the total AoI by jointly optimising scheduling and altitude control.

This work considers a hybrid discrete–continuous action space, where UAV altitude control and IoT device scheduling are closely intertwined with the phase shift configuration of the RIS. This tight coupling introduces significant challenges in designing a control policy that can jointly manage altitude, communication scheduling, and RIS tuning. Directly combining all three components into a unified action space leads to a combinatorial explosion in complexity, making standard DRL algorithms ineffective due to their difficulty handling large discrete action sets. To mitigate this, the proposed control framework focuses on UAV altitude and IoT device scheduling as the primary decision variables, while RIS phase shifts are adaptively configured based on the selected device and its activation pattern.

Within this framework, the RL agent, located at the base station, executes two coordinated actions at each time slot: adjusting the UAV's altitude and scheduling a single IoT device. The RIS phase shifts are then dynamically tuned to support the selected transmission. To learn an effective policy under uncertain and dynamic IoT activation patterns, the framework employs a model-free deep reinforcement learning algorithm based on Proximal Policy optimisation (PPO). This approach enables adaptive control of UAV altitude, communication scheduling, and RIS configuration, aiming to minimise the expected sum AoI.

2.2.6.2 Example 2

This example [21] presents the first work to investigate UAV-assisted IoT networks where UAVs are equipped with virtual queues to enhance data freshness. The network consists of resource-constrained IoT devices that periodically sample a stochastic process, generating time-stamped status-update packets for transmission to a base station (BS). Due to the limited transmission capabilities of these devices, direct one-hop communication to the BS is often unreliable or infeasible. To address this, UAVs function as mobile relays with virtual queues, temporarily storing and retransmitting undelivered data to the BS, thereby improving reliability and reducing information staleness. The UAVs operate over unreliable wireless channels, and their altitude plays a critical role in influencing the propagation conditions between IoT devices and UAVs, as well as between UAVs and the BS. Higher altitudes increase the likelihood of establishing LoS links but result in greater path loss and weaker signals. In contrast, lower altitudes offer stronger signals but less favorable LoS conditions. Given the variability of environmental factors, such as building density and height, dynamic UAV altitude control becomes essential.

In this setup, multiple UAVs are deployed, each assigned to serve a specific cluster of IoT devices. Each IoT device communicates with the nearest UAV, which then relays the data to the BS. UAVs operate in half-duplex mode, meaning they can either receive data from IoT devices or transmit to the BS in a given time slot, but not both. To prevent interference, orthogonal transmission is adopted within each cluster, while inter-cluster interference is mitigated by allocating distinct frequency bands and treating any residual interference as background noise. The core problem is to jointly optimize: (i) the scheduling policy for transmissions from IoT devices to UAVs, (ii) the scheduling of retransmissions from UAVs to the BS, and (iii) the UAVs' altitude control, aiming to minimise the Expected Weighted Sum AoI (EWSA). The decision-making process is sensitive to real-time channel conditions and UAV altitude. To solve this, the control policy is modeled as an MDP, and a PPO algorithm, a robust and stable model-free DRL method, is employed to learn an effective policy. The UAV dynamically determines which data streams to prioritize for relaying, ensuring that the freshest information reaches the BS despite varying channel and environmental conditions.

Thus, the authors formulate the scheduling and UAV altitude adjustment problem as an MDP, with the PPO algorithm deployed to find the optimal policy. Figure 2.11

provides a conceptual reference relevant to the studied scenario. The MDP is formulated as follows:

- **States**: The state s_t at time step t is defined as a combination of the AoI vector at that time step $\mathbf{A}(t)$, the status of the UAV's virtual queue $\boldsymbol{\beta}(t)$, the buffer status vector $\mathbf{b}(t)$, the achievable rate vector $\mathbf{S}(t)$, and the status update size vector that could be sent to the Base Station (BS) at slot t, $\mathbf{S}_U(t)$. Mathematically, $s_t = (\mathbf{A}(t), \boldsymbol{\beta}(t), \mathbf{b}(t), \mathbf{S}(t), \mathbf{S}_U(t))$, where all vectors except $\mathbf{S}_U(t)$ are of the same size, M.
- **Actions**: At each time step t, the UAV executes an action $a_t = (\boldsymbol{\xi}(t), \boldsymbol{\phi}(t), \kappa(t))$. Here:
  - $\boldsymbol{\xi}(t) = (\alpha_{t1,1}, \ldots, \alpha_{t1,M})$, where $\alpha_{t1,i} \in \{0, 1\}$ indicates whether IoT device i is scheduled to transmit a status-update,
  - $\boldsymbol{\phi}(t) = (\alpha_{t2,1}, \ldots, \alpha_{t2,J})$, where $\alpha_{t2,j} \in \{0, 1\}$ determines whether virtual queue j is scheduled to transmit its status update to the Base Station at time-slot t.
  - $\kappa(t) \in \mathcal{H} = \{(1, 0), (0, 1), (0, 0)\}$ represents UAV altitude control, where:
    - $(1, 0)$ indicates an upward adjustment (i.e., +10 m),
    - $(0, 1)$ indicates a downward adjustment (i.e., -10 m),
    - $(0, 0)$ indicates hovering (no altitude change).
- **Reward**: The immediate reward is the sum of penalties incurred when status updates are collected from IoT devices with high AoI values and when the UAV violates altitude constraints. The first penalty is the sum of AoIs across all IoT devices. The second is incurred if the UAV's altitude $z_u(t)$ goes outside the allowable range $[h_{\min}, h_{\max}]$. The agent is trained to minimise these penalties by learning effective scheduling decisions and altitude control strategies.

This work adopts a policy-based DRL framework to solve the formulated joint optimisation problem. A learning agent, deployed onboard the UAV, interacts with the BS and IoT devices by executing a sequence of actions over time. Through continuous observation of environmental feedback and reward signals, the agent incrementally learns an effective policy that jointly governs UAV altitude control and scheduling decisions. By leveraging the PPO algorithm, the agent aims to optimize long-term performance by minimising the AoI. To address the hybrid discrete-continuous nature of the action space, altitude levels are discretized, and the action space is constructed to include combined scheduling and altitude control decisions. This design enables the agent to make informed choices within a manageable action space while steadily improving its policy through repeated interaction with the environment.

2.2.6.3 Example 3

In this example [22], a UAV-RIS-assisted wireless communication system is presented for a dense urban environment, where tall buildings obstruct direct communication between the BS and IoTDs. To overcome the limitations of static RIS deployments, which often fail to adapt to dynamic environments with mobile IoTDs, a UAV equipped with a RIS is introduced. The UAV-RIS dynamically establishes virtual communication links between the BS and obstructed IoTD clusters, enhancing signal propagation and data transmission reliability. IoTDs are grouped into clusters of varying sizes, where each cluster experiences NLOS conditions due to urban blockages. A UAV-RIS is deployed at the network edge to serve as a mobile passive relay, improving coverage and link quality. A ground control station (GCS) is responsible for monitoring the UAV-RIS's location, gathering activation and spatial information from IoTDs, and orchestrating IoTD scheduling and RIS phase shift control.

A generalized RIS phase shift model is introduced to optimize the SNR received at the IoTDs by jointly considering RIS orientation and the number of reflective elements. The objective is to minimise the network-wide average sum AoI by jointly optimising the 3D positioning of the UAV-RIS, the RIS phase shift configuration, and the transmission scheduling of IoTDs. The studied scenario captures key challenges such as dynamic user activation, urban propagation effects, and RIS control in large-scale IoT networks. This setup represents a significant advancement in AoI-aware wireless system design by addressing secure beamforming, throughput maximization, and freshness guarantees through DRL-based optimisation. Overall, a single-UAV-aided data collection problem is considered, based on an architecture involving a RIS-enabled drone BS connected to the terrestrial BS via a backhaul link, with IoT devices blocked by obstacles (see Fig. 2.10). While the figure generally reflects this setup, in this case, the link between the UAV-BS and the ground BS is specifically the backhaul link, representing a variation of the original scenario. The problem is formulated as an MDP to minimise the average sum of AoI (ASoA) across all IoT devices in the UAV-RIS-assisted network. This is achieved by jointly optimising the RIS phase shifts, UAV-RIS positioning, and IoTD scheduling. The example involves formulating an MDP in a continuous state space setting. The definitions of state, action, and reward are provided below:

- **States**: In each time step t, the state $s_t \in \mathcal{S}$ is represented by the vector:

$$s_t = \{A_1(t), A_2(t), ..., A_K(t), \gamma_1(t), \gamma_2(t), ..., \gamma_K(t), z_u(t)\}$$

where $A_k(t) \in \mathbb{N}$ is the AoI of IoT device k at the start of the t-th time step, for $k \in \{1, ..., K\}$ and $z_u(t)$ is the altitude of the UAV. Also, $\gamma_K(t)$ is the SNR received by IoTD K. The entire uplink transmission from the k-th IoT device to the BS is considered for computing the AoI, as shown below:

$$A_k(t+1) = \begin{cases} 1 & \text{if } \gamma_k(t) \geq \gamma_{\min},\ I_k(t) = 1 \text{ and } W_k(t) = 1 \\ A_k(t) + 1 & \text{otherwise} \end{cases}$$

Here, $W_k(t) = 1$ if the k-th IoT device is active at time t, and $I_k(t) = 1$ indicates that the k-th IoT device is scheduled to transmit at time t. At the end of time step t, if the BS receives a packet update from IoT device k, then $A_k(t+1) = 1$. Otherwise, $A_k(t+1) = A_k(t) + 1$.

- **Actions**: A fixed action space is used because its size greatly impacts the stability and convergence of the DRL algorithm. At each time step t, the UAV-RIS performs a multidimensional action $a_t \in \mathcal{A}$.
One dimension of this action space is the UAV's altitude adjustment, given by $\Delta z_u(t) \in \{V_{\max} \cdot \Delta t_{slot}, 0, -V_{\max} \cdot \Delta t_{slot}\}$. These values represent upward movement, hovering, and downward movement, respectively. Note that the UAV-RIS transitions between altitudes with a fixed maximum velocity of $V_{\max}$ (m/s). Also, the UAV-RIS can travel at most $V_{\mathrm{Max}} \times \Delta t_{slot}$ (m)s in each time step.
The second dimension involves a *"stay or switch" approach* for scheduling the IoT device clusters. If cluster k_m is scheduled at time t, the agent can choose to either schedule the same cluster again (stay) or switch to a different cluster at time $t+1$.
The UAV-RIS's decision at each time step t is represented by a multidimensional action a_t. This action has two key dimensions:

1. Scheduling Decision ($I(t)$): A binary variable $I(t)$ dictates the cluster scheduling. If $I(t) = 0$, the UAV-RIS continues to download data from the same IoT device cluster as in the previous time step. If $I(t) = 1$, it switches to downloading from a different cluster.
2. Altitude Adjustment ($\Delta z_u(t)$): This dimension controls the UAV's vertical movement, with three possible values: $\Delta z_u(t) \in \{V_{\max} \cdot \Delta t_{slot}, 0, -V_{\max} \cdot \Delta t_{slot}\}$, representing upward movement, hovering, and downward movement, respectively.

Combining these two dimensions, the action at time t is defined as $a_t = \{I(t), \Delta z_u(t)\}$. Since $I(t)$ has 2 possibilities and $\Delta z_u(t)$ has 3 possibilities, the overall action space $\mathcal{A}$ contains $2 \times 3 = 6$ possible actions in total.
Two criteria determine which cluster to switch to when $I(t) = 1$. Assuming $k'_m \in [1, K_M]$ is the cluster served in the previous time step $t-1$, the first criterion (Criteria 1) identifies the cluster with the highest average AoI:

$$v = \arg\max(A),$$

where $A = \{A_{k_m}(t)\}_{k_m=1}^{K_M}$ is the average AoI of IoT devices in cluster k_m for $k_m \in [1, K_M]$, and $A' = \{A_{k_m}(t)\}$ for $k_m \in [1, K_M]\backslash\{k'_m\}$. IoT device cluster v is scheduled in the current time step if $|v| = 1$. Otherwise, if multiple clusters have the same maximum average AoI, the second criterion (Criteria 2) is used. This criterion selects the cluster with the highest average SNR among those with the same AoI, defined as:

$$v' = \arg\max_{k_m \in v}(\boldsymbol{\gamma}(t)),$$

where $\boldsymbol{\gamma}(t) = \{\overline{\gamma}_{k_m}\}$ is the average SNR among IoT devices in cluster $k_m \in v$. The scheduled IoT device cluster will then be $n_c = v[v']$, meaning $I_{n_c} = 1$ for $n_c \in [1, K_M]$.

Invalid actions are removed from the action space in certain states. For example, if the UAV-RIS is at its maximum height (z_{max}) or minimum height (z_{min}), the actions involving upward or downward movement are eliminated. This allows the agent to focus on optimising the scheduling strategy rather than learning the altitude range of the UAV-RIS.

- **Reward**: The reward function is designed to achieve the objective of minimising the average sum AoI of all IoT devices and is computed at the end of each time step. The immediate reward function is:

$$r_t = -\sum_{k=1}^{K} A_k(t).$$

To address this problem, DQN and PPO algorithms are employed with the objective of learning an optimal control policy that jointly manages the altitude adjustment of the UAV-RIS and the transmission scheduling of IoT devices (IoTDs). These model-free DRL frameworks are designed to adapt to dynamic network environments by effectively coordinating UAV positioning and communication scheduling to minimise the AoI.

2.2.6.4 Summary

This section analyses three works focused on optimising scheduling that involved UAV altitude control, with RIS being a prominent technology deployed in most of these works. RIS with UAVs is critical in dense urban environments where tall buildings obstruct direct communication between the base station and IoT devices. The RIS-carrying UAV improves signal quality and extends communication coverage to the IoT devices. Thus, the main objective in these works is to jointly optimize the UAV's altitude or position, communication scheduling, transmission scheduling of IoT devices, and RIS phase shifts. The problem is formulated as an MDP to minimise AoI across all IoT devices in the UAV-RIS-assisted network. A summary of the states, actions, and rewards considered in this scheduling and altitude control optimisation section is provided in Table 2.6.

2.2.7 Scheduling Optimisation and Packet Drop Rate

In this case, we provide one example, considering it is not a commonly studied problem compared to others.

Consider a UAV-assisted IoT data collection scenario, where SNs sample environmental information at either fixed or random rates and cache the sampled data using

Table 2.6 Summary of MDP modeling for scheduling optimisation and altitude control

Ref.	States	Actions	Rewards
[20]	Vector of AoI, SNR vector from RIS-assisted signal combination, UAV altitude	UAV altitude adjustment	Total AoI
[21]	AoI vector, virtual queue status, buffer status vector, achievable rate vector	IoT device scheduling, virtual queue transmission, UAV altitude control	Penalties from collecting high AoI updates and altitude violations
[22]	AoI of IoT device, UAV altitude	UAV altitude adjustment, stay/switch cluster scheduling	Average sum AoI of all IoT devices

a sample-and-replace policy. An energy-constrained rotary-wing UAV is deployed to collect cached data from each SN as it flies over them. The system comprises a BS, multiple SNs, and a single UAV operating within a defined region of interest. Each SN is equipped with a buffer to store sampled data, such as temperature or air pollution levels. The UAV acts as a mobile relay, forwarding collected data from the SNs to the BS. The target area is partitioned into equally sized square grids, each small enough to contain only one SN. Time is slotted, and in each time slot, the UAV traverses one grid and collects data from the SN if present in that grid.

The UAV-assisted data collection process is structured into two sequential phases. In the first phase, the SN transmits one update packet to the UAV over a LoS link. In the second phase, the UAV forwards the received packet to the BS. The objective is to jointly optimize the SNs' sampling and queuing processes, along with the UAV's relaying strategy, to determine a trajectory that minimises the average AoI while ensuring a low packet drop rate. Thus, in [4], the aim is to formulate the AoI-optimal UAV data collection policy over a finite horizon for IoT devices, subject to the UAV's residual energy, based on an architecture similar to the conceptual model in Fig. 2.5, involving a single UAV. The problem is modeled as an MDP in which the network is divided into n equal-sized grids, and the UAV moves through one grid per time step to collect data from the SN located there. The MDP formulation is detailed below:

- **States**: The state at time step t includes the AoI of each of K SNs, the lifetime of the SNs' update packet, the UAV's horizontal location $q_u(t) = (x(t), y(t)) \in \Omega$ (a projection of the UAV's trajectory on the ground, where Ω contains the center coordinates of all grids, and the UAV can only fly to one of the four adjacent grids in each time step), and the UAV's residual energy $e(t) \in \mathcal{E} \triangleq [0, E_{\max}]$, where $E_{\max}$ is the maximum energy capacity. A state is terminal when the UAV's residual energy is below a threshold for the current slot [4].

- **Actions**: The action space consists of the UAV's flight in each time step from its current grid to one of the four adjacent grids. Thus, the flight action $a_t \in V \triangleq$ {North (N), South (S), West (W), East (E)} represents the UAV's flight direction.
- **Reward**: The cost function is modeled as the weighted sum of the SNs' average AoI, packet drop rate, and energy consumption, as represented in Eq. 2.16, where a_t is the action taken at state s_t, $C(s_t, a_t)$ is the cost function with weighting factor δ_1 for the packet drop rate and δ_2 for energy consumption. $A_k(t)$ represents the AoI of SN k in time step t, K is the total number of SNs, $m_k(t)$ is a binary variable indicating if an update packet is dropped by SN k in time step t, and $e_c(t)$ is the energy consumption in time step t.

$$r_t = C(s_t, a_t) = \frac{1}{K}\sum_{k=1}^{K}(A_k(t) + \delta_1 \cdot m_k(t)) + \delta_2 \cdot e_c(t). \tag{2.16}$$

The finite MDP formulation incorporates the effects of sampling rates, buffer management, and UAV-assisted relaying. To determine the age-optimal UAV trajectory, the MDP is solved using a DQN algorithm, which enables the learning of an asymptotically optimal policy without requiring prior knowledge of the environment dynamics. This DQN-based approach significantly outperforms conventional greedy strategies in simulations by achieving lower AoI and packet drop rates, while also minimising UAV energy consumption. The algorithm proposed to solve the formulated problem is provided in Example 8 under *Value-based Solutions and Algorithms* in the next chapter.

2.3 Partially Observable MDP-Based Modeling and Problem Formulation

POMDPs are not as common as traditional MDPs in the current studies on AoI minimisation for UAV-assisted IoT, and thus, only three examples are provided under this section that considered energy optimisation. Given that the environment offers only partial information during the decision-making process, POMDP-based problems are considerably more challenging. For instance, in one of the works, each UAV operates as an intelligent agent with only partial observations of its location, energy level, local AoI status, and the time-to-live updates from nodes within its coverage. Due to the limitation of having access only to local observations, the problem is commonly modeled as a Decentralised Partially Observable Markov Decision Process (Dec-POMDP), where agents collaborate to minimise a global cost, in this case, the AoI. Figure 2.12 represents an architecture relevant to this section. A summary of the states, actions, and rewards considered for energy optimisation in POMDP-based modeling is provided in Table 2.7.

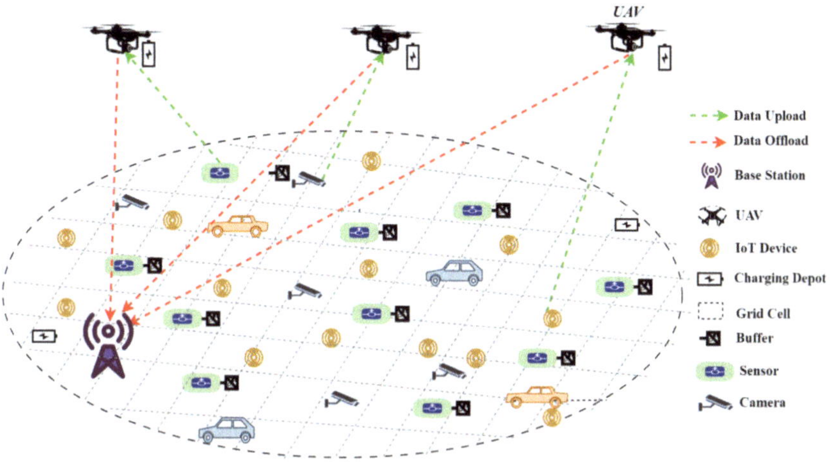

Fig. 2.12 Multi-UAV-assisted IoT network model

Table 2.7 Summary of partially observable MDP modeling for energy optimisation

Ref.	Observation Space	Actions	Rewards
[23]	AoI values, lifetime of SN update packets, UAV's position, UAV's remaining energy	UAV flying directions	Weighted sum of: SNs' average AoI, UAVs' energy consumption, positive constant penalty
[24]	AoI values of SNs, lifetime of SN update packets, UAVs' position vector, UAVs' remaining energy	UAV flying directions	Weighted sum of: SNs' average AoI, UAV energy consumption, penalty term
[25]	AoI values, UAVs' current position, energy state	Hover in place or fly in one of four directions	Weighted average AoI at the base station
[26]	UAV ground projection, AoI of all SNs, difference between UAV's remaining time and time needed to reach final location	UAV movement, SN scheduling	Sensor nodes' weighted average AoI

2.3.1 Example 1

In this example [23], UAVs serve not only as relays but also as energy sources, enabling simultaneous wireless information and power transfer for SNs. These SNs use harvested RF energy to upload sensed information updates to UAVs, thereby improving data freshness and prolonging their operational lifetime. Each SN samples environmental data at either fixed or random intervals, depending on its designated sampling mode. Multiple UAVs cooperate to wirelessly charge the SNs and collect update packets while flying between grids at a fixed altitude. The collected data is subsequently forwarded to a data center. The goal is to minimise the AoI of collected data while accounting for UAV energy consumption and promoting SN longevity. Each UAV operates as an intelligent agent with only partial observations of the environment, which include its own location, energy level, local AoI status, and the time-to-live of the most recent update from each SN within its coverage. To address the limitations imposed by partial observability, the problem is modeled as a POMDP under a time-slotted, grid-based system abstraction.

Thus, UAVs cooperatively collect update packets from ground SNs and relay the data to the data center, based on an architecture similar to the reference model shown in Fig. 2.12, involving multiple UAVs. The objective is to jointly minimise the AoI of the SNs and the energy consumption of UAVs. The multi-UAV data collection problem is formulated as a collaborative multi-agent RL framework, in which each UAV acts as an agent, learns decision-making policies from its interactions with the environment, and receives a joint reward. Each UAV obtains only local observations, and therefore, the problem is reformulated as a Dec-POMDP. In this setting, agents cooperatively minimise a global cost based on their individual partial observations. The POMDP is formulated as follows:

- **Observation space**: At time step t, the observation space of UAV u includes AoI values, $\hat{A}_u(t)$, the lifetime of SN update packets observed by the UAV, $\hat{U}_u(t)$, the UAV's position, $\mathbf{p}_u(t)$, and the remaining energy of the UAV, $E_u(t)$. Thus, the observation space at slot t for UAV u is $z_u(t) = (\hat{U}_u(t), \hat{A}_u(t), \mathbf{p}_u(t), E_u(t))$. The joint observation space for all M UAVs at time t is $z_t = (z_1(t), z_2(t), ..., z_M(t)) \in \mathcal{Z}$. The u-th UAV's location is updated via Eq. 2.17:

$$\mathbf{p}_u(t+1) = \begin{cases} \mathbf{p}_u(t) + (0, L) & \text{if} f_u(t) = \text{North} \\ \mathbf{p}_u(t) + (0, -L) & \text{if } f_u(t) = \text{South} \\ \mathbf{p}_u(t) + (-L, 0) & \text{if } f_u(t) = \text{West} \\ \mathbf{p}_u(t) + (L, 0) & \text{otherwise (East)} \end{cases} \tag{2.17}$$

- **Actions**: Each UAV transmits energy while hovering. Hovering is also done to receive data, after which the UAV flies to an adjacent grid. The possible flying directions are North (N), South (S), West (W), and East (E), i.e., $\hat{\mathcal{V}} \triangleq \{\text{N, S, W, E}\}$. The flight action of each UAV u at time step t is $f_u(t) \in \hat{\mathcal{V}}$. The joint action of all M UAVs at time t is $a_t = (f_1(t), f_2(t), ..., f_M(t))$.

- **Reward**: The cost depends on a weighted summation of the SNs' average AoI, a factor ξ representing the importance of the UAVs' average energy consumption, and a positive constant penalty $p_{penalty} = P_f(t)$ (if an inappropriate action, e.g., leading to a collision, is taken in slot t, or zero otherwise). The reward function is given by Eq. 2.18:

$$C(z_t, a_t) = \frac{1}{K}\sum_{k=1}^{K} A_k(t) + \xi \frac{1}{M}\sum_{u=1}^{M} E_u(t) + P_f(t), \tag{2.18}$$

  where M is the total number of UAVs and u is the index for a UAV. The location of the k-th sensor node is $\Omega_k = [x_k, y_k] \in \mathbb{R}^2$ for $k = 1, ..., K$. Also, $A_k(t) \in \mathcal{A} = \{1, 2, ..., A_{k,\max}\}$ denotes the AoI of sensor node k at the end of slot t, and $A_{k,\max}$ is the maximum AoI of the k-th SN.

The formulated multi-agent reinforcement learning problem considers each UAV as an independent agent that interacts with the environment to make its own decisions while receiving a joint reward. Due to the limitation of only having access to local observations, the problem is modeled as a Dec-POMDP, where agents collaborate to minimise a global cost based on incomplete local information.

Traditional single-agent algorithms become ineffective in large-scale networks due to the complexity of the multi-agent environment. Given the linear structure of the reward function, the value decomposition network (VDN) is adopted as a suitable solution approach. A VDN algorithm is developed based on the DQN architecture, where the joint action-value function is assumed to be additively decomposable into local agent-wise action-value functions. This allows each UAV to learn optimal policies from its own observations while contributing to the optimisation of a shared global objective.

2.3.2 Example 2

This example [24] closely resembles the previous one, involving multiple energy-constrained UAVs deployed for data collection from SNs. Unlike earlier scenarios, the generation times of sensing samples are unpredictable, making UAV scheduling more difficult without prior knowledge of whether each SN adopts a periodic or random sampling strategy. As a result, the UAV trajectory optimisation problem is formulated to minimise AoI, subject to UAV energy limitations and collision avoidance constraints. The AoI-aware data collection problem is examined under two scenarios: full and partial observation. In the full observation case, the problem is addressed using the State-Action-Reward-State-Action (SARSA) algorithm, while the partial observation case is tackled using the VDN framework. The network comprises a single BS, multiple SNs equipped with local buffers, and several rotary-wing UAVs that collect sampled data and relay it to the BS. The monitored region is divided into a grid, with each square assumed to accommodate one SN. UAVs fly from one grid center to another at a fixed altitude, with collision avoidance enforced by prohibiting multiple UAVs from occupying the same grid in the same time slot. Each UAV is assigned the full communication bandwidth to maintain reliable LoS links with SNs. If a UAV has sufficient energy and encounters a grid containing an SN with data, it collects the data and forwards it to the BS. If no SN is present or the buffer is empty, the UAV proceeds to a neighboring grid. Once its energy drops below a defined threshold, the UAV returns to the BS to recharge.

Overall, a multi-UAV-aided data collection problem is considered, based on a network architecture similar to that shown in Fig. 2.12, involving multiple UAVs. The problem is formulated as a POMDP, with the state and action space, state transition dynamics, and long-term expected cost. As for the state transition, the UAV's location updates based on four possible movements: $(0, L)$, $(-L, 0)$, $(L, 0)$, and $(0, -L)$, where L is the grid square's side length. The cost is a weighted sum of the average AoI from sensor nodes, UAV energy consumption, and penalties for invalid actions like collisions, to minimise the discounted return.

- **Observation Space**: At time step t, the observation is composed of the following:

$$s_t = \left(\hat{A}(t), \hat{U}(t), \mathbf{p}_u(t), \eta_u(t)\right),$$

  where $\hat{A}(t)$ represents the AoI values of the SNs, $\hat{U}(t)$ is the lifetime of the SN update packets, $\mathbf{p}_u(t)$ denotes the current position vector of UAV u at time t, and $\eta_u(t)$ is the remaining energy of UAV u. The state update process takes into account whether a particular SN has been served, successful data transmission, and the evolution of AoI and packet lifetimes. The UAV's position is updated using the following rule:

$$\mathbf{p}_u(t+1) = \begin{cases} \mathbf{p}_u(t) + (0, L) & \text{if the UAV moves North,} \\ \mathbf{p}_u(t) + (0, -L) & \text{if the UAV moves South,} \\ \mathbf{p}_u(t) + (-L, 0) & \text{if the UAV moves West,} \\ \mathbf{p}_u(t) + (L, 0) & \text{if the UAV moves East.} \end{cases}$$

- **Actions**: Each UAV hovers to collect data from the sensor nodes and then moves to one of the adjacent grid positions in the next time slot. The possible flight directions are North (N), South (S), East (E), and West (W). The flight action of UAV u at time t is denoted by $A_u^t \in \{\text{N, S, E, W}\}$. Invalid flight actions, such as those leading to out-of-bound positions or collisions, are penalized.
- **Reward**: The reward is modeled as a weighted sum of the average AoI of the SNs, the energy consumption of the UAVs, and a penalty term. The objective is to find an optimal policy to minimise the expected discounted return. The reward function is defined as:

$$C(s_t, a_t) = \frac{1}{K}\sum_{k=1}^{K} A_k(t) + \delta \cdot \frac{1}{M}\sum_{u=1}^{M} \eta_u(t) + P_f(t),$$

where $A_k(t)$ is the AoI of the k-th SN at time t, $\eta_u(t)$ represents the remaining energy of UAV u at time t, δ is a weight that balances the importance of energy consumption, and $P_f(t)$ is a penalty term that is non-zero when the UAV takes an invalid action, such as a collision.

The multi-UAV-aided data collection problem is formulated as a Dec-POMDP due to the limited sensing coverage of each UAV. The objective is to model the system as a cooperative multi-agent reinforcement learning problem, where each UAV acts as an agent and collects the AoI values and the lifetimes of update packets from SNs within its coverage. To solve this problem efficiently, centralised training with decentralised execution (CTDE) is adopted, as it allows agents to share information during training while making independent decisions during execution. This approach is generally more effective than independent learning in multi-agent environments.

Among CTDE methods, VDN is employed, which is built on the DQN framework and well-suited for discrete state and action spaces. Although policy gradient methods are often preferred for continuous domains, they are not used in this case due to the discrete nature of the problem. Instead, VDN learns a linear combination of local action-value functions, which aligns with the structure of the problem. Moreover, the cost function is defined as a linear weighted sum of average AoI, UAV energy consumption, and penalty terms. This linearity further supports the applicability of VDN to this setting.

2.3.3 *Example 3*

This example [25] considers a multi-UAV-assisted IoT network, where UAVs cooperatively collect update packets generated by IoT devices and continuously transmit them to the BS to improve information freshness, measured by the AoI. The update arrivals at the IoT devices are stochastic and unknown, requiring UAVs to plan their trajectories without prior knowledge of these arrival rates. Consequently, the problem is modeled as a Dec-POMDP. The objective is to design distributed UAV trajectories that minimise AoI while ensuring collision avoidance and adhering to energy constraints. The network features several CSs where UAVs can recharge as needed. Each CS is connected to the BS via a high-speed backhaul link, allowing UAVs to offload collected packets during recharging.

This problem is particularly challenging due to: (1) the stochastic nature of update arrivals at IoT devices; (2) limited knowledge of environmental parameters, such as packet arrival rates; and (3) the need for strict collision avoidance, requiring UAVs to maintain safe separation. The network comprises static IoT devices, UAVs, CSs, and a BS, all with known locations. As in earlier examples, the region of interest is partitioned into equal-area square grid cells, and time is slotted into intervals of fixed duration. Update arrivals are assumed to be independent across time slots, with generated packets stored in device buffers, replacing any outdated information. UAVs fly at fixed altitudes and speeds, and both propulsion and communication energy consumption are considered, though the latter is negligible compared to propulsion. Within this framework, UAVs collaboratively plan their trajectories to balance data collection, energy constraints, and the need to return to CSs for recharging. During recharging, UAVs utilize the backhaul link to transmit cached packets to the BS. The ultimate objective is to optimize UAV flight paths in a cooperative manner to minimise the network-wide AoI.

Overall, this example considers a multiple-UAV architecture, where UAVs collect data packets generated by IoT devices in a cooperative fashion and transmit them to the BS. The problem is formulated as a multi-UAV distributed cooperative dynamic trajectory planning problem as a Dec-POMDP, where IoTD update arrivals are stochastic and unknown to the UAVs. The Dec-POMDP is defined as the tuple $(\mathbb{S}, \mathbb{O}, \mathbb{A}, R(\cdot, \cdot))$:

- **States** $\mathbb{S}$: At the beginning of time step t, the joint state of all agents is $\mathcal{S}(t) = \{s_1(t), s_2(t), \ldots, s_M(t)\}$, where the AoIs at IoT devices and the BS, the current position, and the energy state of the u-th UAV are all included in the state of the u-th agent, i.e.,
$$s_u(t) = \{Z(t), B(t), L_u(t), E_u(t)\},$$
where $Z(t) = (Z_{D_1}(t), Z_{D_2}(t), \ldots, Z_{D_K}(t))$ and $B(t) = (B_{D_1}(t), \ldots, B_{D_K}(t))$ denote the AoI values at IoT devices and the BS, respectively. Moreover, $L_u(t)$ is the u-th UAV's current location, and $E_u(t) \in [0, E_{\max}]$ is its current battery state. The state space for all states is denoted by $\mathbb{S}$.
- **Observation Space** $\mathbb{O}$: Let $\mathcal{O}(t) = \{o_1(t), o_2(t), \ldots, o_M(t)\}$ represent the joint observation at time step t. The joint observation is for all agents, and each agent only observes the AoI at one IoT device if it visits that device. In this case, at the beginning of time step t, the observation of agent u is defined as
$$o_u(t) = \{\hat{Z}_u(t), B(t), L_u(t), E_u(t)\},$$
where $\hat{Z}_u(t) = (\hat{Z}^u_{D_1}(t), \hat{Z}^u_{D_2}(t), \ldots, \hat{Z}^u_{D_K}(t))$. If an update packet is successfully transmitted from IoT device D_k to UAV u in the previous time step $t-1$, then $\hat{Z}^u_{D_k}(t) = Z_{D_k}(t-1) + 1$. Otherwise, $\hat{Z}^u_{D_k}(t) = Z_0$, where the constant Z_0 helps distinguish it from valid battery states. The space of all possible observations is denoted by $\mathbb{O}$.
- **Action Space** $\mathbb{A}$: In each time step, a UAV can either keep hovering or fly in any of the four directions: North, South, East, or West. Thus, the joint actions of all agents are represented as $\mathcal{A} = \{\mathcal{A}_u\}_{u \in \mathcal{U}}$, where $\mathcal{A}_u = \{\text{N, S, E, W, H}\}$, for all $u \in \mathcal{U}$.
- **Reward Function** $R(\cdot, \cdot)$: $R(\cdot, \cdot)$ maps the joint state-action pair $\mathcal{S} \times \mathcal{A}$ to a real number. It is the reward obtained by selecting the joint action $\mathcal{A} \in \mathbb{A}$ at joint state $\mathcal{S} \in \mathbb{S}$. The reward involves two components. In each time step t, the cost is the weighted average AoI at the BS, i.e.,

$$c(t) = \sum_{k=1}^{K} \varphi_{D_k} B_{D_k}(t+1),$$

where φ_{D_k} represents the importance weight associated with IoT device D_k. A penalty r_p is applied when the distance safety constraint is violated by executing the joint action $\mathcal{A}(t)$. Thus, the reward can be expressed as

$$R_u(\mathcal{S}(t), \mathcal{A}(t)) = -c(t) + r_p \cdot \mathbb{I}(\text{distance safety violated}),$$

where the indicator function $\mathbb{I}(\cdot)$ equals 1 if the condition is true (i.e., the safety constraint is violated), and 0 otherwise. Considering distance safety, improvement of AoI of update packets, and optimal dynamic trajectory planning policy, π^* maximises the discounted long-term cumulative reward:

$$\pi^* = \arg_\pi \max \lim_{T\to\infty} \mathbb{E}\left[\sum_{t=1}^{T} \gamma^{t-1} R(\mathcal{S}(t), \mathcal{A}(t)) \mid \mathcal{S}(1)\right],$$

where the initial state and discount factor are denoted by $\mathcal{S}(1)$ and $\gamma \in [0, 1)$, respectively. The discount factor γ determines the importance of the current reward while ensuring the long-term cumulative reward is finite.

The distributed trajectory planning problem addressed in this work is formulated as a Dec-POMDP, capturing the AoI dynamics at IoT devices, UAVs, and the BS, while incorporating distance-based safety constraints. To solve this problem, a customized multi-agent deep reinforcement learning (MADRL) algorithm is developed. Specifically, the proposed approach integrates QMIX with gated recurrent unit (GRU) techniques to enable decentralised policy learning under partial observability. This algorithm, referred to as QMAUTP, enables UAVs to cooperatively optimize their trajectories in a distributed manner. The effectiveness of QMAUTP is validated through extensive simulations, demonstrating its superior performance compared to baseline MADRL algorithms.

2.4 Summary

This chapter covers the usage of MDPs in modeling the UAV-assisted WSN/IoT applications. MDPs model sequential decision-making problems, and it is a fundamental bedrock widely used to develop RL-based problems, including minimising AoI in UAV-assisted WSN/IoT applications. Researchers have formulated their problems in various ways in order to improve the AoI, that is, by optimising either trajectory, energy, scheduling, altitude, reducing packet drop, or combinations of them. These presented problems can be categorised as either fully observable MDPs or partially observable MDPs. The problems are considered fully observable MDPs, the agents, usually the UAVs, can fully observe the states, and there are no hidden states and controls. On the contrary, POMDPs involve states that cannot be fully observed by these agents. Various studies have been discussed under these two broad categories, specifically their objectives, and the respective states, actions, and rewards are highlighted. Only a few published works formulate their problems as POMDP problems.

References

1. R.S. Sutton, A.G. Barto, *Reinforcement Learning: An Introduction*, 2nd edn. (MIT Press, Cambridge, MA, A Bradford Book, 2018)
2. N.C. Luong, D.T. Hoang, S. Gong, D. Niyato, P. Wang, Y.-C. Liang, D.I. Kim, Applications of deep reinforcement learning in communications and networking: a survey. IEEE Commun. Surv. & Tutor. **21**(4), 3133–3174 (2019)
3. X. Xiang, S. Foo, Recent advances in deep reinforcement learning applications for solving partially observable markov decision processes (pomdp) problems: Part 1–fundamentals and applications in games, robotics and natural language processing. Mach. Learn. Knowl. Extr. **3**(3), 554–581 (2021)
4. P. Tong, J. Liu, X. Wang, B. Bai, H. Dai, Deep reinforcement learning for efficient data collection in UAV-aided internet of things, in *2020 IEEE International Conference on Communications Workshops (ICC Workshops)*, (IEEE, 2020), pp. 1–6
5. Q. Dang, Q. Cui, Z. Gong, X. Zhang, X. Huang, X. Tao, AoI oriented UAV trajectory planning in wireless powered IoT networks, in *2022 IEEE Wireless Communications and Networking Conference (WCNC)*, (IEEE, 2022), pp. 884–889
6. L. Shi, X. Zhang, X. Xiang, Y. Zhou, S. Sun, Age of information optimization with heterogeneous uavs based on deep reinforcement learning, in *2022 14th International Conference on Advanced Computational Intelligence (ICACI)*, (IEEE, 2022), pp. 239–245
7. A. Ferdowsi, M.A. Abd-Elmagid, W. Saad, H.S. Dhillon, Neural combinatorial deep reinforcement learning for age-optimal joint trajectory and scheduling design in UAV-assisted networks. IEEE J. Sel. Areas Commun. **39**(5), 1250–1265 (2021)
8. C. Zhou, H. He, P. Yang, F. Lyu, W. Wu, N. Cheng, X. Shen, Deep RL-based trajectory planning for AoI minimization in UAV-assisted IoT, in *2019 11th International Conference on Wireless Communications and Signal Processing (WCSP)*, (IEEE, 2019), pp. 1–6
9. X. Wei, G. Zhang, Z. Han, Satellite-controlled uav-assisted iot information collection with deep reinforcement learning and device matching, in *2022 7th International Conference on Intelligent Computing and Signal Processing (ICSP)*, (IEEE, 2022), pp. 1254–1259
10. X. Li, J. Li, D. Liu, Energy-efficient UAV trajectory design with information freshness constraint via deep reinforcement learning. Mob. Inf. Syst. **2021**, 1–9 (2021)
11. X. Li, B. Yin, J. Yan, X. Zhang, R. Wei, Joint power control and UAV trajectory design for information freshness via deep reinforcement learning, in *2022 IEEE 95th Vehicular Technology Conference: (VTC2022-Spring)*, (IEEE, 2022), pp. 1–5
12. S.F. Abedin, M.S. Munir, N.H. Tran, Z. Han, C.S. Hong, Data freshness and energy-efficient UAV navigation optimization: a deep reinforcement learning approach. IEEE Trans. Intell. Transp. Syst. **22**(9), 5994–6006 (2020)
13. E. Eldeeb, D.E. Pérez, J.M. de Souza Sant'Ana, M. Shehab, N.H. Mahmood, H. Alves, M. Latva-Aho, A learning-based trajectory planning of multiple UAVs for AoI minimization in IoT networks, in *2022 Joint European Conference on Networks and Communications & 6G Summit (EuCNC/6G Summit)*, (IEEE, 2022), pp. 172–177
14. M.A. Abd-Elmagid, A. Ferdowsi, H.S. Dhillon, W. Saad, Deep reinforcement learning for minimizing age-of-information in UAV-assisted networks, in *2019 IEEE Global Communications Conference (GLOBECOM)*, (IEEE, 2019), pp. 1–6

15. J. Xu, X. Jia, Z. Hao, Research on information freshness of uav-assisted iot networks based on ddqn, in *2022 IEEE 5th Advanced Information Management, Communicates, Electronic and Automation Control Conference (IMCEC)*, vol. 5 (IEEE, 2022), pp. 427–433
16. M. Sun, X. Xu, X. Qin, P. Zhang, AoI-energy-aware UAV-assisted data collection for IoT networks: a deep reinforcement learning method. IEEE Internet Things J. **8**(24), 17275–17289 (2021). https://doi.org/10.1109/JIOT.2021.3078701.
17. N. Zhang, J. Liu, L. Xie, P. Tong, A deep reinforcement learning approach to energy-harvesting UAV-aided data collection, in *2020 International Conference on Wireless Communications and Signal Processing (WCSP)*, (IEEE, 2020), pp. 93–98
18. L. Liu, K. Xiong, J. Cao, Y. Lu, P. Fan, K.B. Letaief, Average AoI minimization in UAV-assisted data collection with RF wireless power transfer: a deep reinforcement learning scheme. IEEE Internet Things J. **9**(7), 5216–5228 (2021)
19. M. Yi, X. Wang, J. Liu, Y. Zhang, R. Hou, Deep reinforcement learning for energy-efficient fresh data collection in rechargeable UAV-assisted IoT networks, in *2023 IEEE Wireless Communications and Networking Conference (WCNC)*, (IEEE, 2023), pp. 1–6
20. M. Samir, M. Elhattab, C. Assi, S. Sharafeddine, A. Ghrayeb, Optimizing age of information through aerial reconfigurable intelligent surfaces: a deep reinforcement learning approach. IEEE Trans. Veh. Technol. **70**(4), 3978–3983 (2021)
21. M. Samir, C. Assi, S. Sharafeddine, A. Ghrayeb, Online altitude control and scheduling policy for minimizing AoI in UAV-assisted IoT wireless networks. IEEE Trans. Mob. Comput. **21**(7), 2493–2505 (2022). https://doi.org/10.1109/TMC.2020.3042925.
22. M. Sherman, S. Shao, X. Sun, J. Zheng, Optimizing AoI in UAV-RIS-assisted IoT networks: off policy versus on policy. IEEE Internet Things J. **10**(14), 12401–12415 (2023)
23. Z. Li, J. Liu, L. Xie, X. Wang, M. Jin, A deep reinforcement learning approach for multi-uav-assisted data collection in wireless powered iot networks, in *2022 14th International Conference on Wireless Communications and Signal Processing (WCSP)*, (IEEE, 2022), pp. 44–49
24. Z. Li, P. Tong, J. Liu, X. Wang, L. Xie, H. Dai, Learning-based data gathering for information freshness in UAV-assisted IoT networks. IEEE Internet Things J
25. K. Chi, F. Li, F. Zhang, M. Wu, C. Xu, Aoi optimal trajectory planning for cooperative uavs: a multi-agent deep reinforcement learning approach, in *2022 IEEE 5th International Conference on Electronic Information and Communication Technology (ICEICT)*, (IEEE, 2022), pp. 57–62
26. M. Yi, X. Wang, J. Liu, Y. Zhang, B. Bai, Deep reinforcement learning for fresh data collection in UAV-assisted IoT networks, in *IEEE INFOCOM 2020-IEEE Conference on Computer Communications Workshops (INFOCOM WKSHPS)*, (IEEE, 2020), pp. 716–721
27. X. Fan, M. Liu, Y. Chen, S. Sun, Z. Li, X. Guo, RIS-assisted UAV for fresh data collection in 3d urban environments: a deep reinforcement learning approach. IEEE Trans. Veh. Technol. **72**(1), 632–647 (2022)

References

10. J. Xu, X. Ma, Z. Hao, Research on information freshness [illegible]

[illegible]

Chapter 3
Reinforcement Learning and Deep Reinforcement Learning for AoI-Aware UAV-IoT

3.1 Reinforcement Learning and Deep Reinforcement Learning

Reinforcement Learning (RL) is a machine learning method where an agent interacts with an environment, takes actions, observes results, and adjusts its actions to achieve optimal outcomes. Fundamentally, RL involves the agent and the environment. The agent acts upon the environment and receives feedback in the form of rewards or penalties, which guide its learning process (refer to Fig. 3.1). Actions affect the environment, changing its state, and based on the feedback, whether positive reinforcement or negative penalties, the agent refines its decision-making process. DRL incorporates artificial neural networks to guide the actions of the agent. It combines both artificial neural networks and reinforcement learning techniques.

Through this feedback loop, the agent learns about its environment by balancing exploration (taking novel actions to discover new information) and exploitation (leveraging actions known to yield high rewards), observing the rewards or penalties over time. An RL agent is fundamentally characterized by its policy and its learning algorithm. The policy function dictates the agent's behavior by mapping states to chosen actions. While the policy directly defines action selection, it is often derived from or learned by estimating an action-value function (Q-function). The Q-function estimates the expected long-term return (Q-value) of taking a specific action in a given state. For environments with discrete state and action spaces, this Q-function can be stored in a Q-table, which enables the agent to evaluate the quality of its decisions based on its desired objectives. In scenarios with large or continuous state/action spaces, neural networks or other function approximators are typically used to represent these functions instead of tables.

For environments with large or continuous state and/or action spaces, explicitly storing the Q-function in a Q-table becomes impractical or impossible due to the large number of entries. In such scenarios, the Q-function (or the policy directly) is instead approximated by a function approximator, most commonly a neural network.

O. A. Amodu et al., *Markov Decision Processes and Reinforcement Learning for Timely UAV-IoT Data Collection Applications*, Studies in Computational Intelligence 1220,
https://doi.org/10.1007/978-3-031-97011-5_3

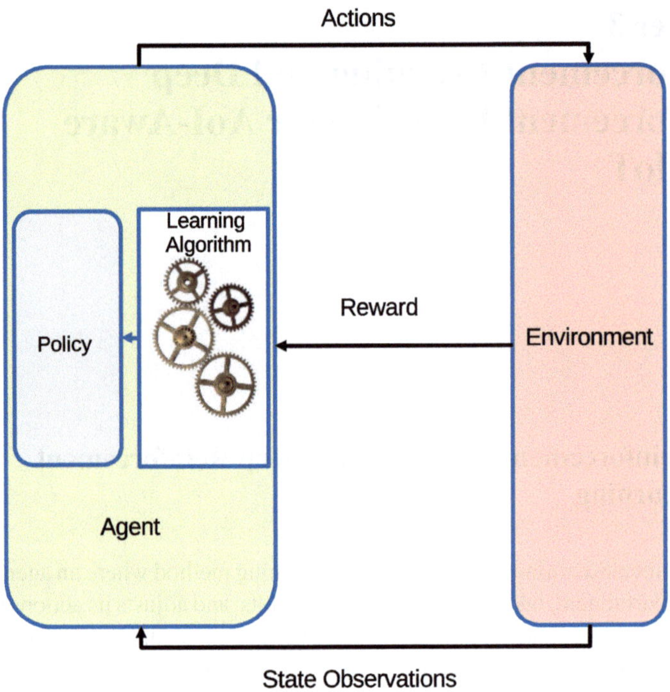

Fig. 3.1 Typical reinforcement learning (RL) architecture

These neural networks form the 'deep' component of DRL and are powerful universal function approximators. This means that, with sufficient architectural capacity and proper training, they can approximate virtually any complex input-output relationship, thereby learning to estimate Q-values or directly represent optimal policies without the need for exhaustive tabular storage.

Neural networks, integral to DRL, represent the policy within the agent, enabling the approximation of complex functions without making the problem overly complex, which would hinder training. In RL, neural networks identify and capture the complexity of the problem, ensuring accurate approximations of state-action spaces.

3.1.1 Bellman Equation

The Bellman equation, a key concept in dynamic programming and RL, provides a recursive decomposition for solving the problem of finding the best policy. It is central to various RL algorithms, including the popular Q-learning.

In particular, the Bellman Equation simplifies the problem of solving the Q-table in Q-learning by breaking it into smaller steps, comparing the value of a state-action pair to the Q-table to obtain the error, which is then used to update the value estimates.

This iterative process continues until the true value of every state-action pair is sufficiently determined, guiding the agent toward the optimal path. The following are Bellman-related equations: (i) the Bellman equation for the state-value function, (ii) the Bellman optimality equation, and (iii) the Bellman equation in Q-learning.

(i) Bellman Equation for State-Value Function

The Bellman equation for the value function $v_\pi(s)$ of a policy π expresses the relationship between the value of the current state s and the expected values of future states. It is written as [1]:

$$v_\pi(s) = \sum_{a \in A(s)} \pi(a|s) \sum_{s',r} p(s',r|s,a)\left[r + \gamma v_\pi(s')\right] \tag{3.1}$$

This equation states that the value of state s under policy π is the expected sum of the immediate reward r and the discounted value of the next state s', weighted by the probability of transitioning from s to s' given action a. Note that and $\gamma \in [0, 1]$ is the discount factor.

(ii) Bellman Optimality Equation

When aiming for an optimal policy, we can express the value of a state under the optimal policy π^* using the Bellman optimality equation:

$$v_*(s) = \max_{a \in A(s)} \sum_{s',r} p(s',r|s,a)\left[r + \gamma v_*(s')\right] \tag{3.2}$$

Here, $v_*(s')$ represents the optimal state-value function. The equation means that the value of state s under the optimal policy is the maximum expected return achievable by selecting the best possible action a.

(iii) Bellman Equation in Q-Learning

In Q-learning, instead of directly learning the value function $v(s)$, the algorithm learns the action-value function $Q(s, a)$, which provides the expected return for taking action a in state s. The Bellman equation for the Q-function is:

$$Q(s,a) = \mathbb{E}\left[r + \gamma \max_{a'} Q(s',a') \mid s,a\right] \tag{3.3}$$

This equation allows us to update the Q-values iteratively. At each time step, the agent updates the Q-value for the current state-action pair (s, a) using the observed reward r and the maximum Q-value for the next state s', discounted by γ:

$$Q(s,a) \leftarrow Q(s,a) + \alpha\left(r + \gamma \max_{a'} Q(s',a') - Q(s,a)\right) \tag{3.4}$$

where:

- α is the learning rate.
- γ is the discount factor.
- r is the immediate reward.
- $\max_{a'} Q(s', a')$ is the highest Q-value for the next state s'.

As previously mentioned, Bellman equation iteratively breaks down the problem into smaller, manageable updates. Hence in Q-learning, by calculating the error between the current estimate of $Q(s, a)$ and the target (the observed reward plus the maximum future reward), the algorithm iteratively refines the Q-table. This error, known as the Temporal Difference (TD) error, is given by:

$$\delta = r + \gamma \max_{a'} Q(s', a') - Q(s, a) \tag{3.5}$$

As this process continues, the true value of every state-action pair is progressively determined, guiding the agent toward the optimal path.

3.1.2 Classes of RL Algorithms

RL algorithms are either model-based or model-free. They can also be classified as value function-based, policy function-based, and actor-critic-based algorithms.

Model-free RL algorithms learn directly from interactions with the environment. The agent needs to explore all possible states, which can make learning inefficient and time-consuming. Model-free RL algorithms address this by learning through trial-and-error, using feedback to update their internal policies or value functions. Conversely, if the environment model is known, model-based RL is preferred. These algorithms learn the dynamics of the environment, predicting outcomes and reducing the need for exploration. Examples include:

- **VPO (Vanilla Policy Optimisation)**: Updates policy parameters using stochastic gradient ascent on policy performance.
- **TRPO (Trust Region Policy Optimisation)**: Takes the largest possible step for improving performance while maintaining the closeness of old and new policies.
- **DDPG (Deep Deterministic Policy Gradient)**: Off-policy algorithm for continuous action spaces, learning a Q-function and a policy using Bellman's equation.
- **TD3 (Twin Delayed DDPG)**: Addresses DDPG's overestimation problem by learning two Q-functions and updating the policy less frequently.
- **PPO (Proximal Policy Optimisation)**: Similar to TRPO but uses first-order methods to ensure new policies are close to old ones.
- **SAC (Soft Actor-Critic)**: Optimizes a stochastic policy in an off-policy manner, balancing expected return and policy randomness (entropy).

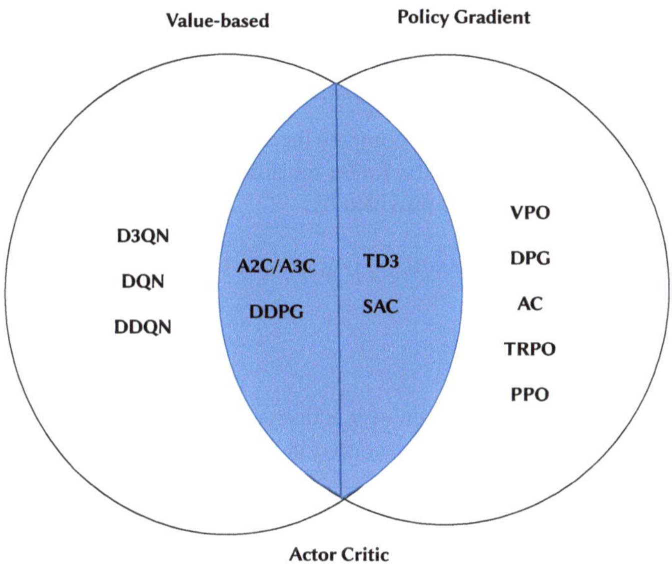

Fig. 3.2 Classification of reinforcement learning algorithms

RL and DRL algorithms are typically grouped into three main types: value-based, policy-based, and actor-critic methods, as shown in Fig. 3.2. In brief, value-based methods focus on learning the value function, policy-based methods aim to learn the policy directly, and actor-critic methods combine elements of both approaches. The choice of which method to use depends on the specific problem and the desired trade-offs between stability, flexibility, and computational efficiency. The following sections will explore these categories in more detail.

3.1.2.1 Value Function-Based Algorithms

Value function-based algorithms evaluate the total discounted rewards from a state, choosing actions with the highest value. The agent updates these values using the Bellman equation, refining decision-making over time. Value functions can handle continuous state spaces with neural networks. However, directly applying value-based methods to infinite (continuous) action spaces is computationally intractable, as it would require maximising over an infinite number of possible actions. Neural networks approximate the value function, enabling the selection of optimal actions. The following explains the value functions learning process in detail.

In value function-based learning, the algorithm takes a state and action as inputs and outputs the value of that action. This value reflects the total discounted rewards expected from that state onward. The policy selects the action with the highest estimated value. In short, the learning process for value functions involves the following steps:

1. Taking random actions.
2. Observing results (states and rewards).
3. Updating values using Bellman's equation.

The Bellman equation serves to compute the value of a state-action pair by taking both the immediate reward and the future rewards from the next state. The update rule for value-based methods is provided by:

$$V(s) = \mathbb{E}_{a\sim\pi}\left[r(s,a) + \gamma V(s')\right] \tag{3.6}$$

where:

- $V(s)$ is the value of state s,
- $\mathbb{E}_{a\sim\pi}$ is the expected value taken over actions a sampled according to policy π,
- $r(s,a)$ is the reward from taking action a in state s,
- γ is the discount factor,
- s' is the next state after taking action a,
- π is the policy.

Deep Q-Network (DQN) [2] stands as a classic example of a value-based algorithm. Figure 3.3 illustrates a generalized DQN architecture wherein the DQN agent interacts with its environment. This example features four state vectors and a deep neural network that aids the agent in learning the optimal policy. The agent executes an action and subsequently receives a reward from the environment.

DQN demonstrates the capability to learn successful policies from high-dimensional sensory inputs through reinforcement learning [2]. The agent engages with the environment via a sequence of observations, actions, and rewards, striving to select actions that maximise the cumulative future rewards. A deep convolutional neural network serves to approximate the optimal action-value function. The algorithm employs experience replay, a mechanism that mitigates correlations within the observation sequence and smooths variations in the data distribution. Furthermore, experience replay stores the agent's experiences, comprising state, action, reward, and next state (s, a, r, s') at each time step, within a **replay memory** encompassing data from different iterations. Q-learning updates are applied to minibatches (or samples) of these experiences, which are drawn uniformly and randomly from the stored pool. Additionally, DQN incorporates an iterative update process that adjusts the action values towards periodically updated target values, thereby reducing correlation with the target.

3.1.2.2 Policy Function-Based Algorithms

Policy function-based algorithms directly learn the policy $\pi(a|s)$ without explicitly estimating the value function, where $\pi(a|s)$ represents the probability of taking action a in state s. This notation aligns with standard policy gradient methods, which optimise the policy directly rather than relying on value estimates. This method,

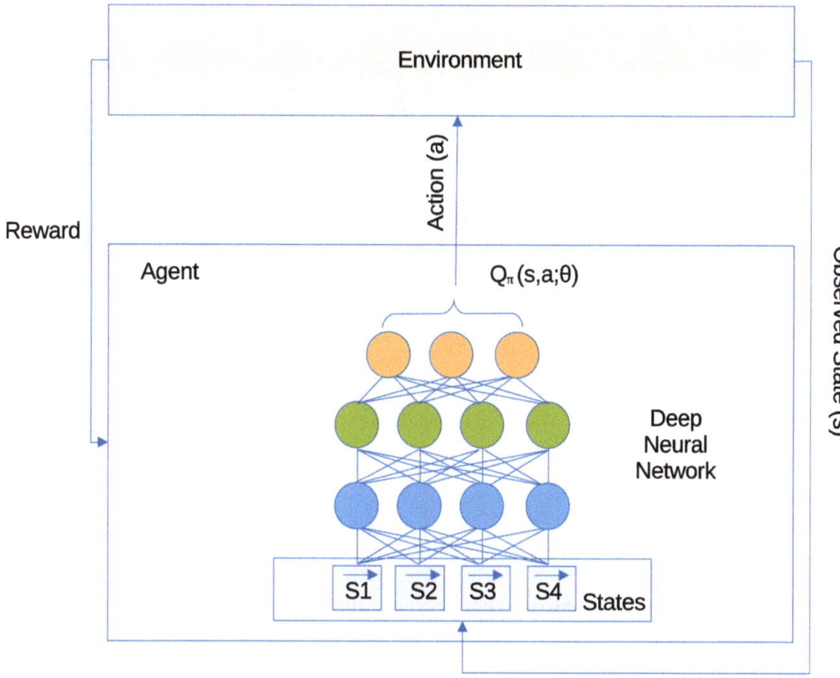

Fig. 3.3 Typical DQN architecture with four state vectors

known as policy gradient, is useful for high-dimensional action spaces. The algorithm of the policy function is as follows:

1. Initialise a random policy $\pi_\theta(a|s)$ parameterised by θ.
2. Select actions using the current policy.
3. Collect experience, observe the state and reward.
4. Evaluate the policy gradient and improve the policy.

No explicit equation like the Bellman equation is used in policy function-based algorithms. Instead, these algorithms optimise the policy $\pi(a|s)$ directly through policy gradients.

Policy gradient methods compute an estimator of the policy gradient, which is plugged into a stochastic gradient ascent algorithm [3]. Generally, policy-based algorithms do not depend on the value function; rather, as shown in Fig. 3.4, it involves a neural network which functions as an actor. PPO [3] is a classic example of a policy-function-based algorithm, although as opposed to others, it also deploys an actor-critic architecture. In other words, traditional policy-based algorithms such as REINFORCE [4] do not involve a critic in their architecture. A typical block diagram of PPO, a policy-based algorithm, is provided in Fig. 3.4, which involves the actor, critic, environment, trajectory, and policy gradient. As the agent interacts with the

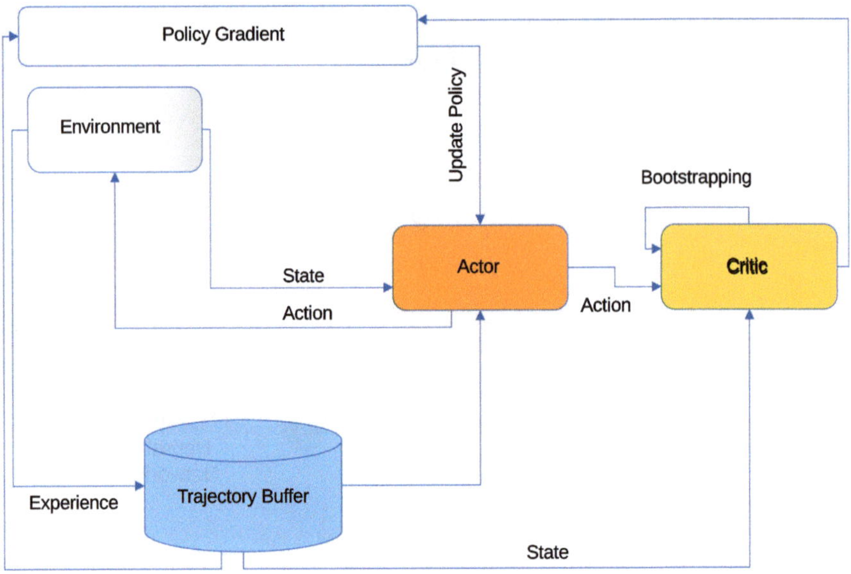

Fig. 3.4 Typical abstraction of proximal policy optimisation (PPO), a policy-based algorithm

environment, it collects a sequence of experiences stored in the trajectory buffer. The policy gradient updates the policy of the actor to obtain the maximum rewards.

3.1.2.3 Actor-Critic Function-Based Algorithms

Actor-critic algorithms combine policy-based and value-based methods. In general, the actor chooses actions, while the critic evaluates them, whereby the critic's feedback helps the actor refine the policy, facilitating learning in continuous action spaces. In specific, these two key components are as follows: (i) actor, which is a neural network that outputs actions based on the current state and is like the policy function in policy gradient methods, and (ii) critic, which is another neural network that estimates the expected future reward (value function) for a given state, and it is similar to the value function in the value-based methods. The following are the equations and steps in actor-critic-based algorithms [5].

(i) Actor or Policy Gradient

Policy ($\pi(a|s)$): Represents the probability of taking action a in state s. The actor serve to maximise the expected return by optimising this policy.

$$\nabla_\theta J(\theta) \approx \frac{1}{N} \sum_{i=0}^{N} \nabla_\theta \log \pi_\theta(a_i|s_i) \cdot A(s_i, a_i) \tag{3.7}$$

where:

- $J(\theta)$ is the expected return under the policy parameterised by θ.
- $\pi_\theta(a|s)$ is the policy function.
- N is the number of sampled experiences.
- $A(s, a)$ is the advantage function:

$$A(s, a) = Q(s, a) - V(s) \tag{3.8}$$

(ii) Critic or Value Function Update

Value Function ($V(s)$): Estimates the expected cumulative reward starting from state s. The critic is the one who models the value function, leading the actor to take actions that yield higher expected returns.

$$\nabla_w J(w) \approx \frac{1}{N} \sum_{i=1}^{N} \nabla_w (V_w(s_i) - Q_w(s_i, a_i))^2 \tag{3.9}$$

where:

- $V_w(s_i)$ is the critic's estimate of the value of state s with parameters w.
- $Q_w(s_i, a_i)$ is the critic's estimate of the action-value of taking action a_i.

(iii) Training Process

1. Initialisation: Initialise actor (θ) and critic (w) parameters.
2. Environment Interaction: Take actions based on the current policy and receive observations/rewards.
3. Advantage Computation: Calculate $A(s, a)$ using current policy and value estimates.
4. Policy Update:

$$\theta_{t+1} = \theta_t + \alpha \nabla_\theta J(\theta_t)$$

5. Value Function Update:

$$w_t = w_t - \beta \nabla_w J(w_t)$$

where: α *and* β are the learning rate for the actor and the critic respectively.

By combining policy and value methods, actor-critic algorithms provide a powerful framework for solving complex reinforcement learning problems, especially those involving continuous action spaces.

In Actor-Critic methods, the actor is the component that implements the policy, while the critic is the component that learns the value function. Actor-only methods, in contrast, operate solely with a policy defined by a set of learnable parameters. For these methods, optimisation procedures are directly applied to the policy parameters, which naturally allows for the generation of continuous actions. On the other

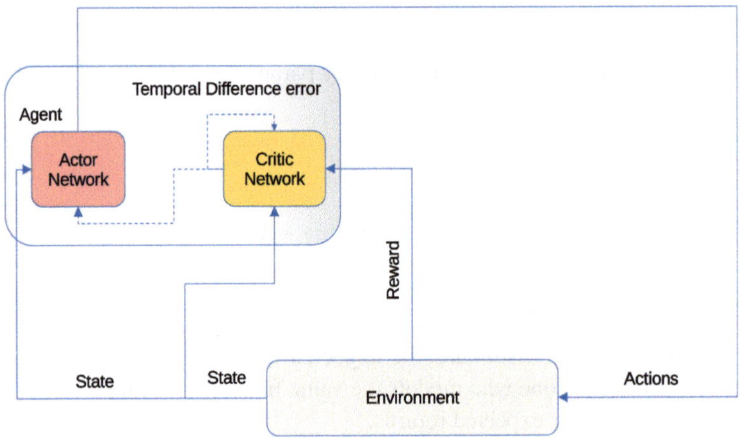

Fig. 3.5 Typical actor-critic architecture in reinforcement learning

hand, the optimisation methods deployed, usually policy gradient methods, have the disadvantage of high variance in estimating the gradients, which slows the learning procedure [6]. Continuous actions can be computed based on the parameterized actor without needing optimisation procedures or a value function. On the other hand, the critic provides the actor with low variance knowledge of the performance. In other words, the estimate of the expected return by the critic allows the actor to be updated with gradients with a lower variance, which makes the learning process much faster [6]. A generic actor-critic framework is provided in Fig. 3.5. Note that in this framework, the actor is the decision maker with a tunable parameter. At the same time, the critic is a function approximator that tries to approximate the value function of the policy deployed by the actor. In contrast, the actor attempts to improve its policy using the current approximation that is provided by the critic, which evolves faster than the actor [7]. The TD error is a feedback signal that trains the actor to improve its policy based on the critic's estimation.

3.1.3 *Multi-Agent Deep Reinforcement Learning*

Multi-agent Deep RL (MADRL) systems involve multiple agents interacting within a shared environment, such as a swarm of UAVs or autonomous vehicles navigating intersections (refer to Fig. 3.6 based on [8]). These agents can be cooperative, working towards a common goal, or adversarial, maximising their benefits while minimising those of others.

In multi-agent systems, defining how agents perform actions and coordinate is crucial. Agents may learn behavior independently or share a common reward function, with each approach having its trade-offs. Shared rewards facilitate collaboration

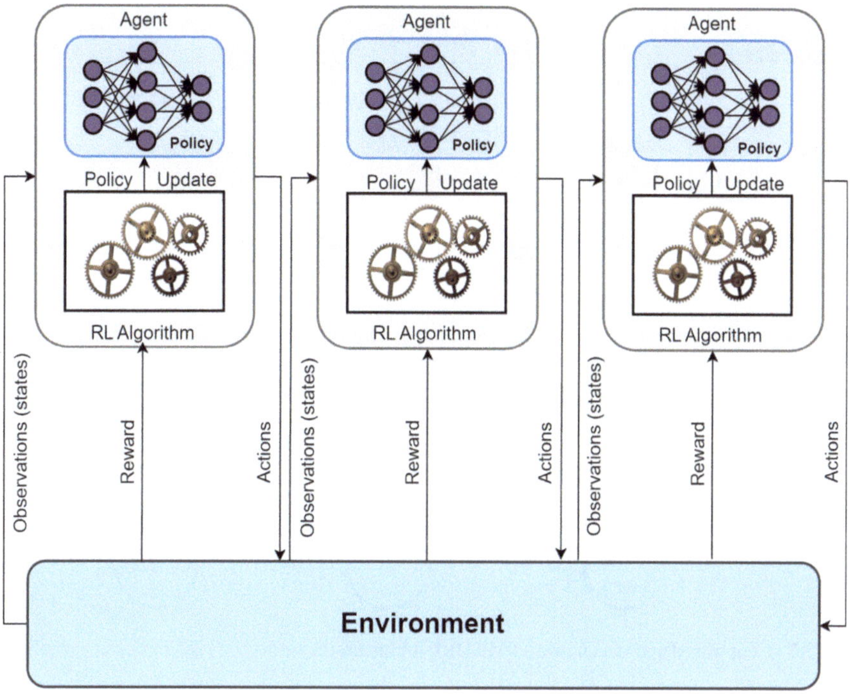

Fig. 3.6 Typical multi-agent deep reinforcement learning (MADRL) architecture with three agents

but may lead to reduced individual contributions from each agent, while localized rewards intensify competition.

In single-agent RL, the goal is to maximise the agent's reward over time. In multi-agent RL, multiple agents interact and learn simultaneously, introducing challenges such as non-stationary environments and the need to update policies over time. Decentralised learning, where each agent is trained independently, minimises communication overhead but may lead to inefficiencies. Centralised learning, where a higher-level process collects and distributes policies, provides a more consistent learning environment (for the centralised trainer) but requires extensive coordination.

In decentralised architectures, agents share experiences minimally, balancing performance and complexity. Centralised architectures, with a higher-level process managing policy learning and distribution, reduce the computational demands on individual agents and provide a more consistent learning environment, facilitating more efficient learning and convergence (refer to Fig. 3.7 based on [8]).

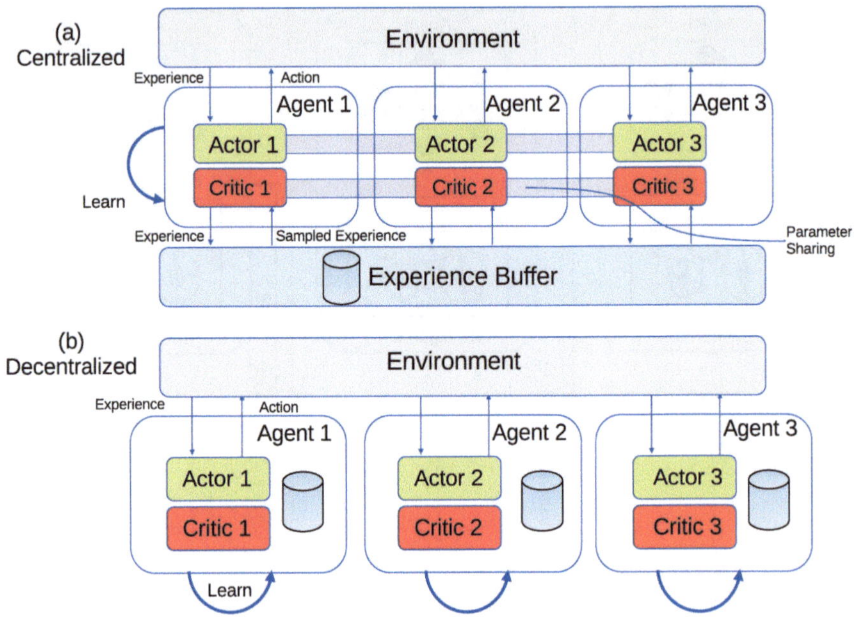

Fig. 3.7 Centralised and decentralised MADRL architectures

3.2 DRL-Based Solutions for AoI-Aware UAV-IoT Applications

This section provides a few selected, detailed DRL-based solutions and algorithms for some of the MDPs highlighted in the previous chapter. These model-free DRL-based solutions are categorised into value-based, policy-based, and actor-critic subsections, respectively. Common notations used in this chapter are summarized in Table 3.1.

3.2.1 Examples of Value-Based Solutions and Algorithms

This section provides examples of value-based algorithms for Age of Information (AoI)-aware UAV-IoT applications (refer Table 3.2). Deep Q-Network (DQN) and Double Deep Q-Network (DDQN) are popular value-based algorithms. A DQN is a multi-layer neural network that outputs a vector of action values for a given state, based on its network parameters. It employs two key components: a target network and experience replay. The target network shares the same architecture as the online network; however, its parameters are copied periodically from the online network (after a fixed number of time steps) and then remain fixed at all other times [9].

Table 3.1 Table of notations

Symbol	Description
α	Learning rate for network updates
α_μ, α_V	Learning rates for actor and critic networks (A3C)
β_1, β_2	Adam optimizer parameters
β	Entropy coefficient (A3C)
γ	Discount factor
ϵ	Exploration rate for an ϵ-greedy policy
ϵ_{decay}	Exploration rate decay factor
$\mathcal{D}$	Replay memory or buffer to store experiences
$\lvert\mathcal{D}\rvert$	Capacity of the replay memory
N_{batch}	Minibatch size for training updates
N_{rep}	Target network update interval
$N_{episodes}$	Number of episodes
t	Time step or time slot
T_{max}, T	Maximum steps in each episode
Z	Delayed update interval (TD3)
s_t	State at time step t
a_t	Action at time step t
r_t	Reward at time step t
y_i	Target value for Q-network update
$\mathcal{L}(\theta)$	Loss function for network training
$\mathcal{L}_{Q_j}$	Critic network loss function (TD3)
θ	General neural network parameters
θ_Q	Online Q-network weights (DDQN)
θ'_Q	Target Q-network weights (DDQN)
θ_μ	Actor network parameters
θ'_μ	Target actor network parameters (TD3)
$\theta_{Q_1}, \theta_{Q_2}$	Online critic network parameters (TD3)
$\theta'_{Q_1}, \theta'_{Q_2}$	Target critic network parameters (TD3)
θ_V	Critic (value) network parameters (A3C)
Φ	Actor network parameters (SAC)
ϕ_1, ϕ_2	Critic network parameters (SAC)
V	Value network (SAC)
V'	Target value network (SAC)
π	Policy (general)
π^*	Optimal policy
$Q(s, a)$	Action-value function
$\hat{A}$	Advantage estimate (PPO)
A_t	Advantage estimate (A3C)

(continued)

Table 3.1 (continued)

Symbol	Description
$\hat{G}$	Estimated returns
$\mathcal{L}_{\text{clip}}$	PPO clipped surrogate objective
$\mathcal{L}_V$	Value function loss
$\mathcal{L}_H$	Entropy bonus loss
ε	Clipping parameter (PPO)
w_j	Worker j (A3C)
α_{temp}	Entropy temperature parameter (SAC)
υ	Soft update rate for target networks (TD3)
τ	Soft update rate for target networks (SAC)
$\mathcal{N}$	Random noise (TD3)
σ	Standard deviation of noise (TD3)
$\widetilde{a}_i$	Noisy target action (TD3)
$\sigma(\cdot)$	Sigmoid function
∇	Gradient operator
b	Bias (neural network)
$\mathcal{W}$	Weight matrix (ANN)
M	Total number of UAVs
$s_u(t)$	State of UAV u at time t
$a_u(t)$	Action of UAV u at time t
$\mathbf{A}_k(t)$	AoI of sensor node k at time t
$\mathbf{A}(t)$	Vector of AoI for all IoT devices at time t
A^*	Minimal/Optimal AoI
E_u	Remaining energy of UAV u
$E_{\max}$	Maximum UAV energy capacity
E_{th}	UAV energy threshold
$\bar{E}_u$	Average UAV energy consumption
E_h	Harvested energy
$\mathbf{p}_u(t)$	Position vector of UAV u at time t
$\mathbf{p}_k$	Position of IoTD k
q_u^*	Optimal UAV trajectory
η	Energy efficiency
η^*	Optimal energy efficiency
b	Scheduling action/policy
b^*	Optimal scheduling policy
p_{drop}	Packet drop rate
ϑ_t	RIS phase shift at time t
ϑ^*	Optimal RIS phase shift control
R	Signal-to-Noise Ratio (SNR)

Table 3.2 Summary of value-based solutions and algorithms based on different optimisation objectives

Refs.	Subsection	Objectives	States	Actions	Rewards and their Constituents
[10]	3.2.1.1	Trajectory	UAV's coordinates, AoI for IoTDs	UAV's speed and flying direction	Negative sum of AoI
[11]	3.2.1.2	Energy	UAV's position and remaining power	UAV's flight	SNs' minimal weighted sum average AoI, minimal UAVs' energy consumption
[12]	3.2.1.3	Trajectory, Energy	UAV's position, energy efficiency, remaining energy, nodes' AoI	UAV's movement	Energy efficiency
[13]	3.2.1.4	Trajectory, Energy	UAV-BS's position, energy, AoI	UAV-BS navigation	Energy efficiency
[14]	3.2.1.5	Trajectory, Energy	UAV's position, IoTD's AoI	UAV's movement	Negative sum of AoI
[15]	3.2.1.6	Trajectory, Scheduling	UAV's location, residual energy, sensors' AoI	UAV's movement	Weighted sum of sensors' average AoI, packet loss rate, energy consumption
[16]	3.2.1.7	Trajectory, Scheduling	UAV's battery level, updated energy level, operational time	UAV's schedule	Reduction in NWAoI value
[17]	3.2.1.8	Scheduling, Packet Drop Rate	UAV's location, energy, sensor's AoI, update packet lifetime	UAV flight direction	SNs' weighted sum average AoI, packet drop rate, energy consumption

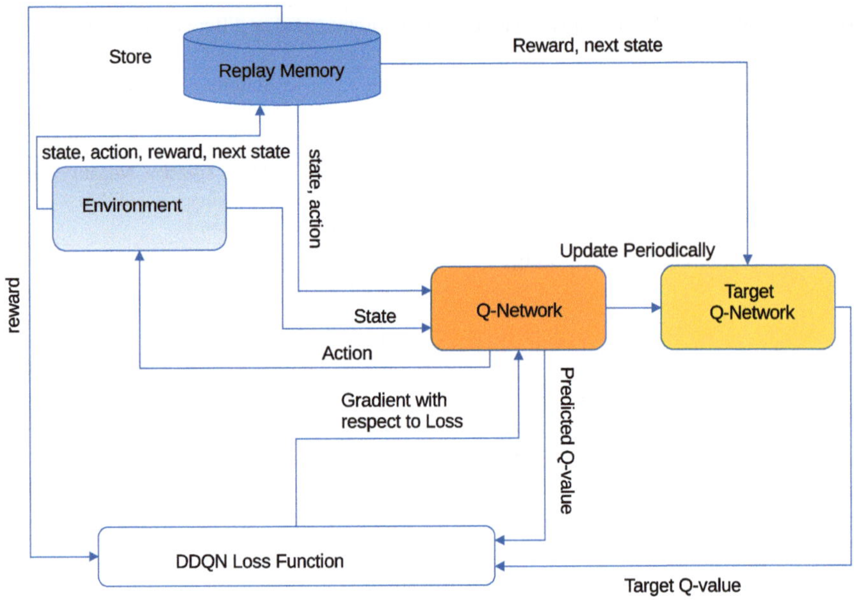

Fig. 3.8 Typical structure of double deep Q-network (DDQN)

DDQN improves upon DQN by mitigating the overestimation of action values. It achieves this by decoupling the max operation in the target value calculation into two distinct steps: action selection and action evaluation. Unlike DQN, DDQN selects actions using the online network (following a greedy policy) but evaluates their values using the target network [9].

An illustration of the DDQN architecture is provided in Fig. 3.8. DDQN was proposed as an enhancement to DQN, which has a tendency to overestimate action values under certain conditions, potentially hindering the algorithm's performance. DDQN applies the principle of Double Q-learning and extends it to function with deep neural networks, leading to more accurate value estimates [9].

3.2.1.1 Example 1: Trajectory Optimisation

Consider a UAV-assisted IoT network deployed in environments lacking cellular coverage, such as remote areas for environmental monitoring or industrial automation [10]. In this setting, a UAV is dispatched to collect data from a massive number of spatially distributed IoT devices by flying over them. These devices generate data following heterogeneous traffic patterns, for instance, periodic traffic from temperature sensors with constant sampling rates, and event-driven traffic from devices like fire alarms that trigger upon threshold violations. To ensure long-range communication and centralised data processing, the UAV is also connected to a satellite. Given the large relative distance between the UAV and the satellite, the satellite-to-UAV link is assumed to be stable, and the data transmission delay to the data center is considered independent of the UAV's trajectory. A high-quality Line-of-Sight (LoS) channel is also assumed between the UAV and IoT devices, enabling successful data collection whenever the UAV is within coverage. In this context, minimising the AoI becomes critical to maintaining fresh data at the data center. However, the UAV's limited flight time and energy constraints, coupled with the dynamic nature of IoT traffic, demand intelligent and adaptive trajectory planning. To model this decision-making process, an MDP framework is adopted. The MDP state captures the UAV's current horizontal and vertical position, as well as the AoI values of all IoT devices. At each decision step, the UAV selects an action comprising its speed and flying direction. The goal is to maximise long-term data freshness, so the reward function is defined as the negative sum of the AoI values across all devices. The environment then transitions to a new state by updating the UAV's location and the AoI values accordingly. To address this, the following example illustrates a learning-based approach that optimizes the UAV's trajectory with respect to AoI in a dynamic IoT environment. This example utilizes the MDP in *Example 1* under *Trajectory Optimisation* in the previous Chapter.

Algorithm 3.1 AoI-based Trajectory Planning (A-TP) Algorithm. Adapted from [10]. Notation and structure were standardized for this book

Require: exploration rate ϵ, decay ϵ_{decay}, discount γ, learning rate α; replay memory $\mathcal{D}$ with capacity $|\mathcal{D}|$; batch size N_{batch}; target update interval N_{rep}; initial UAV position $\mathbf{p}_u(0)$; online Q-net $Q(\cdot; \theta_Q)$ and target Q-net $Q(\cdot; \theta'_Q)$ with $\theta'_Q \leftarrow \theta_Q$
Ensure: optimal UAV trajectory q_u^*

$\mathbf{s}_u(t) \leftarrow \mathbf{p}_u(t)$
while training not converged **do**
  with prob. ϵ: pick random action $\mathbf{a}_u(t)$; else $\mathbf{a}_u(t) \leftarrow \arg\max_{a'} Q(\mathbf{s}_u(t), a'; \theta_Q)$
  execute $\mathbf{a}_u(t)$, observe r_t, $\mathbf{s}_u(t+1)$
  store $(\mathbf{s}_u(t), \mathbf{a}_u(t), r_t, \mathbf{s}_u(t+1))$ in $\mathcal{D}$
  sample minibatch $\{(\mathbf{s}_i, \mathbf{a}_i, r_i, \mathbf{s}'_i)\}_{i=1}^{N_{\text{batch}}} \subset \mathcal{D}$
  for each $(\mathbf{s}_i, \mathbf{a}_i, r_i, \mathbf{s}'_i)$ **do**
    $y_i \leftarrow r_i + \gamma \max_{a'} Q(\mathbf{s}'_i, a'; \theta'_Q)$
    $\mathcal{L}_i \leftarrow \left(y_i - Q(\mathbf{s}_i, \mathbf{a}_i; \theta_Q)\right)^2$
  end for
  update $\theta_Q \leftarrow \theta_Q - \alpha \nabla_{\theta_Q} \frac{1}{N_{\text{batch}}} \sum_i \mathcal{L}_i$
  every N_{rep} steps: $\theta'_Q \leftarrow \theta_Q$
  $\mathbf{s}_u(t) \leftarrow \mathbf{s}_u(t+1)$, $\epsilon \leftarrow \epsilon - \epsilon_{\text{decay}}$
end while
return q_u^*

The aim in this case is to minimise the AoI of IoTD by planning the UAV trajectory and locations based on UAV speed and direction decisions. The proposed algorithm (see Algorithm 3.1) initializes the neural network parameter θ and state s_0. The replay memory $\mathcal{D}$ is also initialised to store the interaction history between the UAV and its environment.

Using the probability ϵ and decay rate ϵ_{decay}, the balance between exploitation and exploration is maintained. Exploration uses a randomised policy, while the exploitation phase uses estimated Q-values to select promising actions. The probability ϵ decays according to the decay rate ϵ_{decay}, leading the agent to exploit learned experiences over time.

Transition information, including selected action, state, reward, and the next state, is fed back into the environment and stored in replay memory. A mini-batch from the memory is sampled to train the neural networks, updating weights and biases. The target value is calculated, and gradient descent with a learning rate α is used to minimise the loss function.

Results show that the proposed algorithm reduces the AoI of data collected by IoT devices compared to benchmark schemes, including a round-robin scheduler with a fixed policy and an age-based max-weight scheduler [18] that relies on current AoI and estimated traffic generation at each IoT device. Additionally, the proposed algorithm demonstrates the ability to adapt to dynamic environments.

3.2.1.2 Example 2: Energy Optimisation

In this example [11], UAVs function not only as relays but also as energy sources, enabling both information and energy transfer to ground-based sensor nodes (SNs). The SNs harvest energy wirelessly from UAVs and use it to transmit their sensed data updates back to the UAVs, thereby improving the freshness of collected information. Depending on their sampling modes, SNs generate updates either at fixed intervals or in response to environmental triggers. Multiple UAVs operate cooperatively in a time-slotted system, flying across a grid-based environment at a fixed altitude, providing RF energy to sensors, and collecting data packets for delivery to the data center. To ensure efficient data freshness and prolong the operational life of SNs, UAVs must adaptively coordinate their movements and energy transmissions based on local, partial observations, such as their own location, residual energy, observed AoI, and the remaining lifetime of update packets from nearby sensors. To address the challenges of limited observability and decentralised decision-making, the problem is formulated as a Dec-POMDP. In this framework, each UAV independently observes its local environment, including the AoI values and update packet lifetimes of nearby sensors, its own position, and its remaining energy. UAVs then choose an action consisting of a directional flight move (North, South, East, or West), during which they can also hover to transmit energy or receive updates. The reward is based on a weighted sum of the average AoI of the SNs, the UAVs' energy consumption, and a penalty if an invalid action, such as one leading to a collision, is taken. Through coordination based on local views, UAVs aim to optimize a global reward, improving network-wide AoI while managing energy efficiency and avoiding collisions. The following example illustrates how a learning-based SARSA algorithm can be applied in this context to achieve decentralised energy-aware data collection. This example utilizes the MDP in *Example 2* under *Energy Optimisation* in the previous Chapter.

Algorithm 3.2 SARSA-based UAV Data Collection Algorithm. Adapted from [11]. Notation and structure were standardized for this book

Require: learning rate α, discount γ, exploration rate $\epsilon_{\min}$, number of episodes N_e, number of steps N_s, max UAV energy $E_{\max}$, energy threshold E_{th}

Initialize $Q(s, a) \leftarrow 0$ for all (s, a)
for ep= 1 to N_e **do**
  Initialize average AoI $\bar{A} \leftarrow 0$, average energy $\bar{E}_u \leftarrow 0$; set $E_m(0) \leftarrow E_{\max}$ for each UAV m; observe s_0
  Choose a_0 via ϵ-greedy on Q
  for $t = 0$ to $N_s - 1$ **do**
    for each UAV m **do**
      if $E_m(t) < E_{\text{th}}$ **then**
        redirect flight $f_m(t)$ toward BS
      end if
    end for
    Execute joint action a_t and observe reward r_t and state s_{t+1}
    Update AoI $A_k(t)$, UAV positions $q_u(t)$, and energy $E_m(t + 1)$
    Choose a_{t+1} via ϵ-greedy on Q
    On-policy SARSA update: $Q(s_t, a_t) \leftarrow Q(s_t, a_t) + \alpha\big[r_t + \gamma\, Q(s_{t+1}, a_{t+1}) - Q(s_t, a_t)\big]$
    Accumulate $\bar{A}$ and $\bar{E}_u$; $s_t \leftarrow s_{t+1}$; $a_t \leftarrow a_{t+1}$
    Decay $\epsilon \leftarrow \max\{\epsilon_{\min}, \epsilon \cdot \epsilon_{\text{decay}}\}$
  end for
end for
return final average AoI $\bar{A}/N_s$ and average UAV energy consumption $\bar{E}_u/N_s$

In this work, multiple energy-limited UAVs are deployed to collect update packets from ground SNs to minimise the average AoI ($\bar{A}$) while adhering to collision avoidance and energy capacity constraints. The authors use a SARSA-based algorithm to address the formulated problem, although applying the SARSA approach becomes highly challenging as the network scales to larger sizes.

For the SARSA-based solution (see Algorithm 3.2), the network parameters are initialised as follows: learning rate α, discount factor γ, minimum exploration rate ϵ_{min}, number of episodes N_e, number of steps N_s, parameter O, maximum UAV energy E_{max}, and positive constants β_1 and β_2. The Q values for time step t ($Q(s_t, a_t)$) are initialised to 0 for all UAV states s_t and actions a_t. Additionally, the initial exploration rate ϵ is set as $\min(\beta_1, \beta_2)$.

The following steps are then executed in a loop, which ends when the counter for the number of episodes (num) is incremented:

- Initialise $t = 0$ (starting from the first time step), UAV energy in step t ($E_m(t) = E_{max}$), average AoI ($\bar{A} = 0$), and average UAV energy consumption ($\bar{E}_u = 0$). Observe the initial system state s_t.
- For each time step t (up to $N - 1$) and each UAV ($m \in M$), perform three steps: – Reselect a flight direction towards the base station (BS) ($f_m(t)$) for UAV m if $E_m(t) < E_{th}$. – Set the UAV's state for data collection based on the state (s_t) and

the SN sampling state $o_k(t)$ for all SNs. – UAVs choose data collection actions a_t and execute the chosen actions.
- Calculate the energy consumption of the UAV and update the packet status for the k-th SN, UAV position $q_u(t)$, AoI of the k-th SN ($A_k(t)$), and the energy of the m-th UAV for the next time step $(t+1)$. Obtain the reward r_t for the action.
- Choose the next action a_{t+1} using the greedy policy. Reshape the next flight direction for each UAV m if $E_m(t+1) < E_{th}$.
- Update the action-value function ($Q(s_t, a_t)$) and calculate the accumulated SN AoI and UAV energy consumption.

VDN was deployed to solve this problem, similar to the solution in [19]. VDN enables the UAV to make independent decisions regarding flight and data collection based on partially observed network information. Results show that the proposed learning-based algorithm outperforms baseline methods, including DQN, greedy, and nearest neighbor algorithms, across various sampling rates.

3.2.1.3 Example 3: Trajectory and Energy Optimisation

In this example [12], we consider a communication system composed of multiple single-antenna UAV base stations, each serving a large number of ground-based single-antenna IoT devices. Operating within a fixed flight altitude and coverage radius, the UAVs act as aerial data collectors, aiming to efficiently gather updates from IoT devices while preserving their own energy reserves. The primary goal is to design UAV trajectories that maximise energy efficiency while also minimising the AoI of the received data, all within a highly dynamic environment. To address the non-convex nature of the problem and environmental uncertainty, the system is modeled as an MDP. In this MDP, each UAV's state includes its current position, remaining energy, energy efficiency, and the AoI of received data. The UAVs choose flight directions from a discrete set of angles (e.g., left, right, forward, backward), and update their positions accordingly. The reward is determined by how well the UAV satisfies the optimisation constraints, with a positive reward assigned when energy efficiency and AoI objectives are met, and zero otherwise. Due to the nonlinear dependencies among energy, mobility, and data freshness, a DQN-based learning algorithm is employed to learn effective trajectory policies. The following example demonstrates how RL can be used to jointly optimize UAV trajectory and energy efficiency in large-scale IoT deployments. This example utilizes the MDP in *Example 1* under *Trajectory and Energy Optimisation* in the previous Chapter.

The work in [12] presents a DQN-based algorithm for achieving energy efficiency maximisation with AoI constraints. In each training episode, the environment and states s_t are initialized, the UAV observes the current state s_t, selects an action a_t (randomly with probability ϵ), executes it, computes the energy efficiency η, residual energy E, and AoI A, and obtains the next Q-value to determine the next state s_{t+1}. The reward r_t is computed, and the experience (s_t, a_t, r_t, s_{t+1}) is stored in the experience replay memory $\mathcal{D}$.

Algorithm 3.3 DQN-Based Trajectory Design for Maximising Energy Efficiency. Adapted from [12]. Notation and structure were standardized for this book

Input: Learning rate α, discount factor γ, initial exploration rate ϵ, decay rate ϵ_{decay}, memory size $|\mathcal{D}|$, batch size N_{batch}, target network update interval N_{rep}, max episodes N_{episodes}, observed time T_{obs}, step counter $N_{\text{step}} \leftarrow 0$
Output: Energy efficiency η^*, trajectory q_u^*, residual energy E^*, AoI A^*
Initialize Q-network $Q(\cdot; \theta_Q)$ and target network $Q(\cdot; \theta_Q^{\text{Target}})$ with weights $\theta_Q^{\text{Target}} = \theta_Q$
for episode = 1 to N_{episodes} **do**
  Initialize environment and observe initial state s
  while not terminal and $t < T_{\text{obs}}$ **do**
    With probability ϵ, select a random action a
    Otherwise, select $a = \arg\max_{a'} Q(s, a'; \theta_Q)$
    Execute a and observe next state s', compute reward r based on η, E, A
    Store transition (s, a, r, s') in experience replay memory $\mathcal{D}$
    If memory size exceeds $|\mathcal{D}|$, discard oldest entry
    $N_{\text{step}} \leftarrow N_{\text{step}} + 1$
    if $N_{\text{step}} \bmod N_{\text{rep}} = 0$ **then**
      Update target network weights: $\theta_Q^{\text{Target}} \leftarrow \theta_Q$
    end if
    Sample mini-batch of size N_{batch} from $\mathcal{D}$
    Compute target Q-value using the Bellman equation: $y = r + \gamma \max_{a'} Q(s', a'; \theta_Q^{\text{Target}})$
    Update Q-network weights θ_Q by minimizing the Bellman loss
    $s \leftarrow s'$
  end while
  Decay exploration rate: $\epsilon \leftarrow \epsilon \cdot \epsilon_{\text{decay}}$
end for
return η^*, q_u^*, E^*, A^*

The estimated Q-network (with weights θ_Q) is duplicated to the target network (with weights θ'_Q) periodically, and the Q-network is trained based on the loss function to optimise θ_Q using a learning rate α. Then the next iteration begins.

For the fixed replay memory $\mathcal{D}$ with size $|\mathcal{D}|$, training data are updated every N_{tr} steps to replace historical data, refreshing the training data in memory. The random sampling method forms the batch (of size N_{batch}) by choosing experiences from replay memory $\mathcal{D}$ randomly, avoiding dependence on the stored data and smoothing changes between historical data and new observations. This way, the best trajectory q_u^*, scheduling strategy b^*, and ultimately energy efficiency η^* and AoI A^* are obtained when training converges. The algorithm input and output are summarised in Algorithm 3.3. Simulation results demonstrate that the proposed scheme, which utilizes this algorithm, achieves better energy efficiency compared to the scheme that uses DQN without experience replay.

3.2.1.4 Example 4: Trajectory and Energy Optimisation

In this example [13], the focus is on designing energy-efficient trajectories for multiple UAVs functioning as mobile aerial base stations, aimed at delivering timely information from a large number of IoT devices. These UAVs operate within fixed sensing ranges and altitudes, and the scenario includes single-antenna UAV base stations and IoT devices commonly used in smart grid, agricultural, safety, or geographic sensing environments. To jointly optimize energy efficiency and data freshness captured via AoI, while respecting constraints like remaining battery, safety distance, and network dynamics, the problem is modeled as an MDP. The state for each UAV includes its current position, target location, average energy efficiency, and the average AoI associated with navigation. The UAV selects actions that move it from one valid position to another, considering communication coverage and trajectory limitations. The reward function incentivizes satisfying energy efficiency, AoI, and path constraints, while penalizing violations accordingly. Given the non-convex nature of the objective and the complexity of real-time navigation in dynamic environments, classic optimisation methods become inadequate. Therefore, a learning-based approach is adopted to find a near-optimal policy. The following example presents a DQN method with experience replay for learning the trajectory policy in this energy-constrained UAV navigation scenario, as shown in Algorithms 3.4 (training) and 3.5 (testing). This example utilizes the MDP in *Example 3* under *Trajectory and Energy Optimisation* in the previous Chapter.

In this work, a navigation policy π^* is designed for multiple UAVs deployed to improve information freshness and IoT device connectivity. The UAV-BS navigation problem is solved via deep Q-learning using a DQN (with weights θ_Q) and a target DQN (with weights θ'_Q). The network parameters θ_Q are initialised randomly.

At each time step t of the training episode, the energy efficiency η metric is calculated while considering both navigation and communication parameters. Using an ϵ-greedy policy, the energy efficiency η is used for calculating the reward r_t function in the exploration phase, and the action a_t is taken from the current DQN using an exploration probability ϵ. Considering the context of AoI A and other constraints, the reward r_t is observed, and the system transitions to the next state s_{t+1}.

In the exploration phase, the UAV-BS can explore joint actions a_t to achieve an improved reward value for choosing the most energy-efficient actions. The mini-batch approach is used, where the experience (s_t, a_t, r_t, s_{t+1}) is shuffled from the replay memory $\mathcal{D}$ buffer randomly to remove the observation sequence, making the distribution of the observation data smoother. Stochastic gradient descent with learning rate α is used to train the DQN, updating the network parameter θ_Q and network bias b.

Results show that this approach improves energy efficiency by 3.6% compared to the greedy algorithm and by 3.13% compared to the baseline DQN approach.

Algorithm 3.4 Training Phase: DQN with Experience Replay for UAV-BS Trajectory Policy Optimisation in Navigation. Adapted from [13].. Notation and structure were standardized for this book

Input: Learning rate α, discount factor γ, initial exploration rate ϵ, decay rate ϵ_{decay}, memory size $|\mathcal{D}|$, batch size N_{batch}, target network update interval N_{rep}, max episodes N_{episodes}, observed time T_{obs}, step counter $N_{\text{step}} \leftarrow 0$
Output: Optimal trajectory policy π^*
Initialize Q-network $Q(s, a; \theta_Q)$ with random weights
Initialize target Q-network $Q(s, a; \theta_Q^{\text{Target}})$ with weights $\theta_Q^{\text{Target}} \leftarrow \theta_Q$
for episode = 1 to N_{episodes} **do**
  Initialize environment and observe initial state s
  while trajectory not terminal **do**
    With probability ϵ, select a random action a
    Otherwise, select $a = \arg\max_{a'} Q(s, a'; \theta_Q)$
    Execute a, observe next state s', and compute reward r from η, E, A
    Store transition (s, a, r, s') in experience replay memory $\mathcal{D}$
    If memory size exceeds $|\mathcal{D}|$, discard oldest entry
    $N_{\text{step}} \leftarrow N_{\text{step}} + 1$
    if $N_{\text{step}} \bmod N_{\text{rep}} = 0$ **then**
      Update target network weights: $\theta_Q^{\text{Target}} \leftarrow \theta_Q$
    end if
    Sample mini-batch of size N_{batch} from $\mathcal{D}$
    Compute target Q-value using the Bellman equation: $y = r + \gamma \max_{a'} Q(s', a'; \theta_Q^{\text{Target}})$
    Update Q-network weights θ_Q by minimizing the Bellman loss
    $s \leftarrow s'$
  end while
  Decay exploration rate: $\epsilon \leftarrow \epsilon \cdot \epsilon_{\text{decay}}$
end for
return Optimal policy $\pi^*(s) = \arg\max_a Q(s, a; \theta_Q)$

Algorithm 3.5 Testing Phase: UAV-BS Trajectory Policy Execution. Adapted from [13]

Input: Trained Q-network $Q(s, a; \theta_Q)$
Output: Final UAV-BS trajectory τ^*
for each time slot $t = 1$ to T **do**
  Observe current state s_t
  Select action: $a_t = \arg\max_a Q(s_t, a; \theta_Q)$
  Update UAV-BS position and trajectory: $\tau^* \leftarrow \tau^* \cup a_t$
  Transition to next state s_{t+1}
end for
return UAV-BS trajectory τ^*

3.2.1.5 Example 5: Trajectory and Energy Optimisation

This example [14] considers the dynamic planning of UAV trajectories to minimise the AoI while accounting for the network's energy constraints and unpredictable wireless environments. IoT devices upload their sensed data using energy harvested from UAVs under practical Line-of-Sight/Non-Line-of-Sight (LoS/NLoS) channel conditions. To address this, a proactive UAV trajectory planning strategy is employed using DRL, allowing UAVs to adapt their flight paths according to dynamic channel variations and the trade-off between energy transmission and data collection. The network is structured into distributed subsystems, where each UAV is responsible for charging and collecting data from IoT devices within its coverage area. To avoid interference, devices are restricted from simultaneously harvesting energy and transmitting data. The underlying decision process is modeled as an MDP, where each UAV's state includes its current location, represented by a hexagonal grid index, and a normalized AoI metric. Actions correspond to movements between adjacent hexagons or remaining stationary, while the reward is defined as the negative sum of AoI values across all devices in the coverage area. The following example introduces a proactive UAV trajectory planning algorithm that leverages DRL to optimize information freshness in energy-constrained IoT networks. This example utilizes the MDP in *Example 4* under *Trajectory and Energy Optimisation* in the previous Chapter.

Algorithm 3.6 AoI-aware Proactive UAV Trajectory Planning Algorithm. Adapted from [14]. Notation and structure were standardized for this book

Input: Initialise Q-network with weights θ, target Q-network with weights θ', and replay memory $\mathcal{D}$
Output: UAV optimal location q_u^*, scheduling b^*, and minimal average AoI
while training not converged **do**
    Observe current state s_t representing UAV location and device AoI status
    With probability ϵ, select a random action $a_t \in \mathcal{A}$
    Otherwise, select $a_t = \arg\max_a Q(s_t, a; \theta)$
    Execute action a_t; observe reward r_t and next state s_{t+1} based on AoI variation
    Store transition (s_t, a_t, r_t, s_{t+1}) in replay memory $\mathcal{D}$
    Sample mini-batch of transitions from $\mathcal{D}$
    for each transition (s_j, a_j, r_j, s_{j+1}) in the batch **do**
        Compute target value: $y_j = r_j + \gamma \max_{a'} Q(s_{j+1}, a'; \theta')$
        Compute loss: $\mathcal{L} = (y_j - Q(s_j, a_j; \theta))^2$
        Update Q-network by minimizing loss via gradient descent
    end for
    Update target Q-network weights: $\theta' \leftarrow \theta$ every O steps
    Update current state: $s_t \leftarrow s_{t+1}$
    Decay exploration rate ϵ
end while
return Optimal UAV location q_u^*, scheduling strategy b^*, and minimal AoI

A DQN-based proactive trajectory planning algorithm was investigated in [14]. The work uses experience replay $\mathcal{D}$, which stores transitions necessary for Q-learning and creates mini-batches for updating the neural networks. DQN adopts experience replay and evaluation and target networks to overcome unstable learning. The minibatch update from the experience replay $\mathcal{D}$ is crucial within the DQN framework for estimating the parameters using the loss function:

$$L(\theta) = \left[r(s_t, a_t) + \gamma \max_{a \in \mathcal{A}} Q(s_{t+1}, a \mid \theta') - Q(s_t, a_t \mid \theta) \right]^2 .$$

The inputs and outputs of the algorithm is provided in Algorithm 3.6, where ϵ helps to balance exploration and exploitation. During the former, the UAV chooses a random action a_t from its action space $\mathcal{A}$ while in the latter, using the probability $1 - \epsilon$, the UAV selects a potentially promising action a_t using a greedy policy in view of the available Q-value estimates.

After the action a_t is taken, the UAV observes the reward r_t and transits to the next state s_{t+1}. Then it stores the experience (s_t, a_t, r_t, s_{t+1}) in the replay memory $\mathcal{D}$, samples a random mini-batch from $\mathcal{D}$, and obtains the target value y. Afterward, gradient descent is performed on $(y_j - -Q(s, a|\theta)^2$, and the target parameters θ' are updated in every finite number (O) of steps. The current state s_t is updated. The agent gradually tends to choose an optimal policy π^* as the probability ϵ reduces or decays.

The proposed DQN-based algorithm achieves a 20% to 65% reduction in AoI compared to three other existing trajectory planning algorithms considered in the evaluation.

3.2.1.6 Example 6: Trajectory and Scheduling Optimisation

This example [15] considers a UAV-assisted IoT network composed of a single base station, a rotary-wing UAV, and multiple SNs, each equipped with a data buffer for storing sampled environmental information. Unlike previous examples, the incorporation of buffers introduces a unique queuing dynamic, where new packets replace older ones. The target monitoring region is discretized into equal-sized grids, and the objective is to determine an optimal UAV flight trajectory that minimises AoI while considering the UAV's residual energy and environmental factors. A finite MDP framework is adopted to model the UAV's sequential decision-making process. During its mission, the UAV must collect data from the sensors while maintaining an energy level above a threshold, below which it is required to return to the base station. The AoI of each sensor node is influenced by a combination of its sampling process, queuing delay, and the UAV-assisted data transmission latency. The MDP's state space captures the AoI and packet lifetime vectors, the UAV's current position, and its remaining energy. The UAV's actions consist of directional moves across adjacent grid cells, while the reward function penalizes high AoI, packet loss, and energy consumption. This example applies DRL via a DDQN algorithm to jointly optimize UAV trajectory planning and data collection scheduling under practical network constraints. This example, utilizes the MDP in *Example 2* under *Trajectory and Scheduling Optimisation* in the previous Chapter.

Algorithm 3.7 DDQN-based UAV-assisted Data Collection Algorithm Adapted from [15]. Notation and structure were standardized for this book

Input: Learning rate α, discount factor γ, exploration rate ϵ, memory capacity $|\mathcal{D}|$, batch size N_{batch}, maximum episodes N_{episodes}, observation period T_{obs}
Output: UAV flight trajectory q_u^*, minimal average AoI, and minimal packet loss rate
Initialise experience replay memory $\mathcal{D}$ with size $|\mathcal{D}|$
Initialise current Q-network with parameters θ and target network with θ'
for episode $= 1$ to N_{episodes} **do**
    Initialise the environment and initial state s_t
    while UAV has sufficient energy **and** $t < T_{\text{obs}}$ **do**
        With probability ϵ, select a random action a_t
        Otherwise, select $a_t = \arg\max_a Q(s_t, a; \theta)$
        Execute action a_t, observe reward r_t, and transition to next state s_{t+1}
        Store experience (s_t, a_t, r_t, s_{t+1}) in memory $\mathcal{D}$
        Sample a mini-batch of transitions from $\mathcal{D}$
        for each transition (s_j, a_j, r_j, s_{j+1}) in the batch **do**
            Determine $a^* = \arg\max_a Q(s_{j+1}, a; \theta)$
            Estimate target value: $y_j = r_j + \gamma Q(s_{j+1}, a^*; \theta')$
            Compute loss: $\mathcal{L}(\theta) = (y_j - Q(s_j, a_j; \theta))^2$
        end for
        Update θ by minimizing $\mathcal{L}(\theta)$ via gradient descent
        if update interval is reached **then**
            Update target network: $\theta' \leftarrow \theta$
        end if
        $s_t \leftarrow s_{t+1}$
        Decay exploration rate ϵ
    end while
end for
return UAV trajectory q_u^*, minimal average AoI, and minimal packet loss rate

The authors in [15] investigate a UAV-assisted, information freshness-aware data collection framework for IoT environments. To address the problem of trajectory optimisation, they propose a DDQN-based UAV-assisted Data Collection Algorithm (see Algorithm 3.7) to derive the optimal UAV flight path q_u^*.

The algorithm begins by initializing key parameters: experience replay memory $\mathcal{D}$ of size $|\mathcal{D}|$, learning rate α, discount factor γ, and exploration rate ϵ (initially set high to encourage exploration). A neural network with parameters θ is initialized, and the number of episodes $N_{episodes}$ is set as an input.

For each episode, the initial state s is established. An action is selected based on the ϵ-greedy policy, which balances exploration and exploitation by comparing a random value with ϵ. As training progresses, ϵ is decayed to increase the likelihood of choosing the optimal action.

The UAV executes the chosen action in state s, observes the resulting reward, and transitions to a new state by updating its location. The resulting experience tuple is stored in the replay memory $\mathcal{D}$ following a random placement strategy.

A mini-batch of transitions of size N_{batch} is sampled randomly from $\mathcal{D}$ to compute the loss function $L(\theta)$. Gradient descent is used to minimise this loss, thereby updating the current value network parameters θ. After a fixed interval N_{step}, the parameters of the target network are synchronized with those of the current network.

The algorithm continues iterating until either the UAV's energy drops below a predefined threshold or the maximum number of episodes $N_{episodes}$ is reached. The outputs of the algorithm include the optimized UAV trajectory q_u^*, AoI of the SNs, and the packet loss rate.

Results show that the algorithm improves the UAV's AoI performance when collecting data from ground nodes and reduces the packet loss rate of SNs. In other words, the DDQN algorithm achieves a lower average AoI compared to both the DQN and baseline greedy algorithms.

3.2.1.7 Example 7: Trajectory and Scheduling Optimisation

This example [16] considers a UAV-assisted data collection scenario where both the UAV and ground IoT devices are constrained by limited energy resources. The UAV collects time-sensitive status updates from energy-constrained SNs with the goal of optimising its flight path and scheduling decisions to minimise the normalized weighted AoI (NWAoI) of different physical processes observed at the UAV. These processes may represent diverse applications such as agriculture, healthcare, industrial monitoring, or public safety. Uplink data transmissions from IoT devices to the UAV are considered, with all transmitted packets assumed to be small in size. The UAV operates at a fixed altitude, and time is divided into discrete slots. Due to the complexity introduced by both trajectory planning and update scheduling under energy constraints, the problem is modeled as a mixed-integer finite-horizon MDP. In this setup, the state captures both the residual battery levels of SNs and their update histories, while the UAV's action space includes the decision to either terminate the scheduling or select a specific node to transmit an update. The reward function reflects the improvement in the NWAoI metric resulting from each action. The following example applies DRL techniques to efficiently learn energy- and freshness-aware scheduling policies under trajectory constraints. This example utilizes the MDP in *Example 3* under *Trajectory and Scheduling Optimisation* in the previous Chapter.

The work in [16] employs DRL for the joint optimisation of UAV trajectory and update status scheduling in UAV-assisted networks with multiple ground nodes. The proposed DRL (see Algorithm 3.8) runs offline before the UAV is deployed and comprises two parts: an ANN to reduce the state-space dimension by extracting necessary features, and RL to find the best policy based on these features. For deriving the policy that maximises the total expected system reward, [16] deploys the DQN architecture.

The Normalised Weighted AoI (NWAoI) minimisation algorithm works as follows: Initialise the replay memory $\mathcal{D}$ storing UAV's previous experiences, the Q-network with weights θ_Q, and set the number of iterations to $k = 1$. The initial time step is set to $t = 1$, an empty policy is initialized, and the NWAoI is set to 1 while observing the initial state s_1.

Algorithm 3.8 Deep RL for NWAoI Minimisation Algorithm. Adapted from [16]. Notation and structure were standardized for this book

Input: Replay memory $\mathcal{D}$, learning rate α, discount factor γ, exploration rate ϵ, Q-network weights θ_Q, target network weights θ_Q^{Target}
Output: Optimal flight trajectory u^*, Normalised Weighted AoI (NWAoI)
Initialise episode counter $k \leftarrow 1$
Initialise initial state s_1, empty policy $u_0 = []$, and NWAoI $G(\cdot) = 1$
repeat
  Set time step $n \leftarrow 1$
  repeat
    With probability ϵ, select a random action a_n
    Otherwise, select $a_n = \arg\max_\alpha Q(s_n, \alpha; \theta_Q)$
    Update policy: $u_n = [u_{n-1}, a_n]$
    Solve the environment model to obtain updated NWAoI $G(u_n)$
    Compute reward: $r_n = G(u_{n-1}) - G(u_n)$
    Observe next state s_{n+1}
    Store transition (s_n, a_n, r_n, s_{n+1}) in $\mathcal{D}$
    $n \leftarrow n + 1$
  until termination condition is met
  Sample a mini-batch of transitions from $\mathcal{D}$
  for each transition (s_j, a_j, r_j, s_{j+1}) in the batch **do**
    Compute target value: $y_j = r_j + \gamma \max_{\alpha'} Q(s_{j+1}, \alpha'; \theta_Q^{\text{Target}})$
    Compute loss: $\mathcal{L}(\theta_Q) = (y_j - Q(s_j, a_j; \theta_Q))^2$
    Update Q-network by minimizing loss via gradient descent
  end for
  Update target network weights $\theta_Q^{\text{Target}} \leftarrow \theta_Q$ periodically
  Decay exploration rate ϵ
  Increment episode: $k \leftarrow k + 1$
until convergence
return Optimal policy u^* and final NWAoI

Select a random action a_t with probability ϵ, otherwise select an action a_t based on the maximum of $\arg\min_a Q(\overline{s}_t, a|\theta_Q)$ where $\overline{s}_t$ is the state representation vector and θ_Q contains the Q-network weights. Then, update the policy with the selected action, solve the NWAoI problem using the updated policy, observe the reward r_t and new state s_{t+1}, and store the experience (s_t, a_t, r_t, s_{t+1}) in replay memory $\mathcal{D}$. Increment t by 1 and repeat the process until the terminal state is reached.

Sample a batch of random experiences from the replay memory $\mathcal{D}$, calculate the target value depending on whether it is in the terminal state or not, derive the gradients of all episodes in the batch, and train the Q network based on the average gradient using learning rate α. Continue until convergence is achieved. For large-scale networks, [16] proposed a long short-term memory-based autoencoder to ensure a fixed-size vector representation for the state space.

The proposed DQN-based algorithm simultaneously reduces the complexity of the system state and learns the optimal policy for scheduling the transmission of status updates from ground nodes.

3.2.1.8 Example 8: Scheduling Optimisation and Minimal Packet Drop Rate

This example [17] constitutes a simplified model that focuses on studying the packet drop rate in a UAV-assisted IoT data collection setting. The system consists of a base station, several SNs, and a rotary-wing UAV operating in a region divided into square grids. Each grid is assumed to contain only one SN, and time is slotted such that the UAV moves across one grid per time slot. SNs sample data, such as temperature or air pollution, at fixed or random rates and cache the sampled updates using a sample-and-replace policy. The UAV, constrained by limited energy, flies over these grids to collect data from the SNs and subsequently relays it to the base station. Each SN is equipped with a buffer for storing sampled data. The data collection process unfolds in two phases: in the first, the SN transmits a single update to the UAV via a LoS link; in the second, the UAV forwards the received update to the BS. The goal is to jointly consider SN data sampling, queue management, and UAV relaying to derive an age-optimal UAV trajectory. To tackle the decision-making complexity, the problem is formulated as a finite MDP, capturing key constraints related to data freshness, packet losses, and UAV energy dynamics. The state space includes the AoI and packet lifetime at each SN, as well as the UAV's horizontal location and residual energy, while for the action, the UAV flies either north, south, west, or east at each time step. This example utilises the MDP under *Scheduling Optimisation and Packet Drop Rate* in the previous Chapter.

The aim in [17] is to minimise the average AoI and packet drop rate in energy-constrained UAV-based IoT data collection. The work deploys DQN to find the most suitable policy π^*. The number of time steps for updating the target network weights θ'_Q is N_s, with N_e representing the number of episodes and ε representing the probability of selecting a random action a_t.

At the end of each episode, the exploration rate $\epsilon = \max\{\beta_1, \beta_2 \cdot \epsilon_{decay}^{num}\}$ is updated, where num is the index of the episode with β_1 and β_2 as two positive constants, and ϵ_{decay} is the decay factor. Note that β_1 serves as a floor for the exploration probability, and β_2 is a scaling factor for the exploration probability that decays over episodes. The capacity of the replay memory $\mathcal{D}$ is represented as G, from which a minibatch of H experiences is extracted to update the weights of the evaluation network θ_Q.

The loss function is calculated as $L(\theta_Q) = \frac{1}{H}\sum_{i=1}^{H}[(y_i - Q(s_i, a_i; \theta_Q))^2]$, where y_i is the target value. Weights θ_Q are updated based on $\theta_Q = \theta_Q - \alpha\nabla_{\theta_Q} L(\theta_Q)$, with gradients calculated from experience:

$$\nabla_{\theta_Q} L(\theta_Q) = \frac{2}{H}\sum_{i=1}^{H}[(y_i - Q(s_i, a_i; \theta_Q))] \left(\nabla_{\theta_Q} Q(s_i, a_i; \theta_Q)\right).$$

Each episode ends when the UAV runs out of battery. The proposed algorithm converges to an asymptotically optimal policy, and results show that it achieves lower AoI and packet drop rate compared to the baseline greedy algorithms.

Algorithm 3.9 DQN-Based UAV-Aided Data Collection. Adapted from [17]. Notation and structure were standardized for this book

Require: init Q-network with weights θ_Q, target network with $\theta_Q^{\text{Target}} = \theta_Q$, replay memory $\mathcal{D}$ with capacity G; learning rate α, discount factor γ, exploration rate ϵ; exploration decay schedule; target network update frequency N_s; batch size H; number of episodes N_e

```
for episode = 1 to N_e do
    reset environment; observe initial state s_0
    while UAV's residual energy e(n) > E_th do
        with probability ϵ select random action a_t; else select a_t = arg max_a Q(s_t, a; θ_Q)
        execute action a_t, observe reward r_t and next state s_{t+1}
        store transition (s_t, a_t, r_t, s_{t+1}) in D
        sample a minibatch of H transitions from D
        for each transition (s_i, a_i, r_i, s_{i+1}) in the batch do
            compute target value: y_i = r_i + γ max_{a'} Q(s_{i+1}, a'; θ_Q^Target) (or y_i = r_i if s_{i+1} is terminal)
            minimize loss (y_i − Q(s_i, a_i; θ_Q))^2 with respect to θ_Q
        end for
        if current step mod N_s = 0 then
            update target network weights: θ_Q^Target ← θ_Q
        end if
        update state: s_t ← s_{t+1}
    end while
    update ϵ ← max{β_1, β_2 · ϵ^{decay(episode)}}
end for
return average AoI Ā and packet drop rate p_drop
```

3.2.2 *Examples of Policy-Based Solutions and Algorithms*

In this section, examples of Policy-based solutions to the AoI minimisation problem in UAV-assisted IoT are provided. Proximal Policy Optimisation (PPO) is a commonly used algorithm in this domain. Refer to Table 3.3 for the summary of the examples of algorithms related to PPO.

PPO is a policy-based actor-critic algorithm that alternates between sampling data through interaction with the environment and optimising a surrogate objective function using stochastic gradient ascent. In contrast to standard policy gradient methods that perform one gradient update per data sample, PPO employs a novel objective function that allows for multiple epochs of minibatch updates during sampling [3]. A detailed block diagram of PPO is provided in Fig. 3.9, which illustrates the interaction between the actor, critic, environment, trajectory, and policy gradient. PPO utilizes a clip function to establish a lower bound on policy performance. This clipped surrogate objective prevents excessively large policy updates by limiting the extent to which the new policy can deviate from the old one. Similarly, PPO uses a loss function that combines both the policy surrogate term and the value function error term. A critic network is used to estimate the advantage, and backpropagation is applied to refine the actions. PPO also involves updating the decisions of both the old actor (the previous policy) and the new actor (the updated policy). The PPO

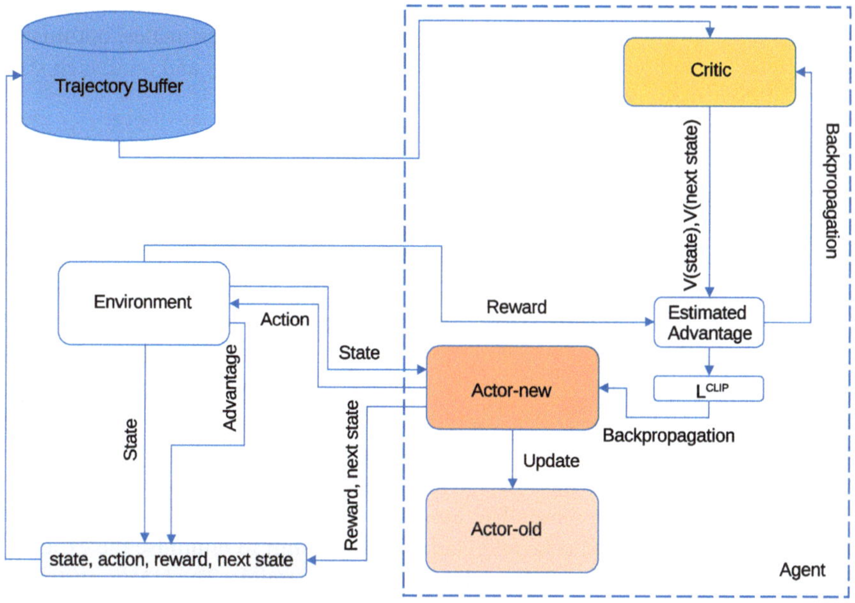

Fig. 3.9 Typical architecture of the PPO policy-based algorithm

Table 3.3 Summary of the PPO-based solutions and algorithms based on optimisation objectives

Refs.	Subsection	Objectives	States	Actions	Rewards
[20]	3.2.2.1	Scheduling, Altitude Control	UAV's altitude, IoTDs' AoI, SNR	Transmission scheduling, UAV's updated altitude	Negative sum of AoI
[21]	3.2.2.2	Scheduling, Altitude Control	IoTDs' AoI, virtual queue status, buffer status, achievable rate, update size status	UAV's altitude	Negative sum of AoI
[22]	3.2.2.3	Scheduling, Altitude Control	UAV's altitude, uplink transmission value, IoTDs' AoI	UAV's movement, cluster activity scheduling	Negative of average sum of AoI

process entails running the current policy in the environment for a certain number of time steps to gather experiences, computing the advantage estimates, optimising the surrogate objective while considering the number of epochs and minibatch size, and finally replacing the old policy with the new policy [3].

3.2.2.1 Example 1: Scheduling Optimisation and Altitude Control

This example [20] introduces reconfigurable intelligent surfaces (RIS) into UAV-assisted IoT data collection to enhance the propagation environment through programmable reflections, thereby improving the desired signal and mitigating interference. RIS technology is both spectrally and energy efficient. The system comprises a single UAV equipped with RIS, which functions as a passive relay forwarding data sampled by IoT devices to the BS. The UAV supports data delivery from IoT devices with unknown activation patterns, and successful data transmission to the BS occurs only if the signal-to-noise ratio (SNR) exceeds a specified threshold–only then is the AoI considered to decrease. The aim is to jointly optimize the UAV altitude, IoT device communication schedule, and RIS phase shift to minimise the expected sum AoI, subject to altitude, SNR, and scheduling constraints. The UAV's altitude is adaptable and managed by the BS within a predefined range, while the BS also computes the RIS phase shifts and controls IoT device scheduling via reliable channels. Although the locations of the BS and IoT devices are known, the device activation patterns are not. The optimisation problem, due to its mixed-integer and non-convex nature, is formulated as an MDP and solved using offline planning before UAV dispatch. This example utilizes the MDP in *Example 1* under *Scheduling Optimisation and Altitude Adjustment* from the previous Chapter. The states encompass the IoT devices' AoI and SNR from the RIS-assisted signal, and the UAV's altitude, while the actions involve transmission scheduling and updated altitude for the UAV (either to increase, decrease, or remain static).

The algorithm proposed in [20] is designed to jointly optimize three key aspects: the UAV's altitude (h_u), the communication schedule (b), and the Reconfigurable Intelligent Surface (RIS) element phase shifts (ϑ). The learning process commences with an initialization step, where the UAV's initial altitude h_0 is set randomly, and the parameters (θ) of the Deep Neural Networks (DNNs) employed in their approach are randomly initialized, ensuring identical network structures.

During each training iteration, the PPO algorithm alternates between a *data sampling phase*, where the agent interacts with the environment for $N_{episodes}$ episodes to collect a batch of experience data, and an *optimisation phase*, where this collected data is used to perform multiple optimisation steps to update the policy parameters.

Within each episode, at every time step t, the current Channel State Information (CSI) is acquired, encompassing links between the IoTDs and the UAV, as well as between the UAV and the Base Station (BS). Subsequently, the optimal phase shifts for the RIS elements (ϑ_t) are determined to achieve a coherent combination of signals from different paths at the BS. Following these adjustments, the agent gathers its observations from the environment, which typically include the Age of

Algorithm 3.10 PPO-Based UAV-RIS Scheduling Framework. Adapted from [20]. Notation and structure were standardized for this book

Require: policy network π_θ, old policy network $\pi_{\theta_{\text{old}}}$, value network V_θ; discount factor γ; clip fraction ε; learning rate α; Adam optimizer; number of episodes N_{episodes}; maximum time steps $T_{\max}$

Initialize DNN parameters θ; set $\theta_{\text{old}} \leftarrow \theta$; set initial UAV altitude h_0
for each training iteration **do**
  Clear episode buffer $\mathcal{B}$
  for episode $= 1$ to N_{episodes} **do**
    Reset environment; observe initial state s_0
    for $t = 0$ to $T_{\max} - 1$ **do**
      Observe AoI vector A_t, SNR vector Υ_t, and UAV altitude $h_u(t)$
      Determine RIS phase shift configuration ϑ_t
      Sample action $a_t \sim \pi_{\theta_{\text{old}}}(\cdot|s_t)$
      Execute action a_t (UAV altitude h_u and communication schedule b), observe reward r_t and next state s_{t+1}
      if UAV altitude constraint is violated **then**
        Apply penalty P to reward r_t
      end if
      Store transition (s_t, a_t, r_t, s_{t+1}) in buffer $\mathcal{B}$
    end for
  end for
  Compute advantage estimates $\hat{A}_t$ and returns $\hat{G}_t$ from the buffered transitions
  for each epoch **do**
    Form minibatches of size N_{mb} from $\mathcal{B}$
    Compute ratio: $r_\theta(s, a) = \frac{\pi_\theta(a|s)}{\pi_{\theta_{\text{old}}}(a|s)}$
    Compute PPO objective: $\mathcal{L}_{\text{PPO}} = \mathbb{E}_t[\min(r_\theta \hat{A}_t,\ \text{clip}(r_\theta, 1-\varepsilon, 1+\varepsilon)\hat{A}_t)] - c_V \mathbb{E}_t[(V_\theta(s_t) - \hat{G}_t)^2] + c_H \mathbb{E}_t[H(\pi_\theta(\cdot|s_t))]$
    Update parameters θ by maximizing $\mathcal{L}_{\text{PPO}}$ using Adam
  end for
  Update old policy: $\theta_{\text{old}} \leftarrow \theta$
end for
return Optimal policy π^*, expected sum of AoI (ESA)

Information (AoI, A_t), the Signal-to-Noise Ratio (SNR, Υ_t) (which reflects the effect of the coherently combined signals at the RIS), and the UAV's current altitude ($h_u(t)$).

The UAV executes an action a_t from the existing policy $\pi_{\theta_{old}}$, where the sampled action is the current UAV altitude h_u and scheduling policy b according to the RIS phase configuration ϑ_t for each IoTD, which achieves maximum SNR. The states s_t of the MDP are defined with a binary value: 1 indicating the IoTD has been scheduled to transmit status updates and 0 otherwise.

Throughout each episode, the RL agent guides the UAV to avoid actions that violate the altitude constraint by applying a penalty P whenever needed. After each episode, the RL trajectory data (s_t, a_t, r_t, s_{t+1}) is buffered for the next phase, and an advantage estimate $\hat{A}$ is computed for efficient training of the policy.

To update the policy network π_θ, the PPO clip objective function in each epoch is computed, and the overall PPO objective function is optimised using the stochastic gradient algorithm with Adam and learning rate α. The old policy $\pi_{\theta_{old}}$ is then set to π_θ, and buffered data is dropped before the new iteration begins.

$$L^{\text{CLIP}}(\theta) = \mathbb{E}_t\left[\min\left(\frac{\pi_\theta(a_t|s_t)}{\pi_{\theta_{old}}(a_t|s_t)}\hat{A}(s_t, a_t),\right.\right. \tag{3.10}$$

$$\left.\left.\text{clip}\left(\frac{\pi_\theta(a_t|s_t)}{\pi_{\theta_{old}}(a_t|s_t)}, 1+\varepsilon, 1-\varepsilon\right)\hat{A}(s_t, a_t)\right)\right], \tag{3.11}$$

In Eq. 3.11, the clip fraction ε controls the clip range. In time step t, the advantage estimate $\hat{A}(s_t, a_t)$ helps to mitigate the large variance of the gradient.

3.2.2.2 Example 2: Scheduling Optimisation and Altitude Control

This example [21] is the first to explore the use of UAVs equipped with virtual queues in UAV-assisted IoT networks. In this setting, low-resource IoT devices periodically sample stochastic processes and aim to deliver timely status updates to a BS. Due to their limited transmission capabilities, one-hop direct transmission to the BS is often infeasible. To overcome this, UAVs act as relays, receiving undelivered data and forwarding it to the BS over unreliable links. The UAV's altitude significantly influences the quality of the propagation channel: higher altitudes increase the likelihood of LoS links but introduce greater path loss, and vice versa. Thus, dynamic altitude control is critical, especially in environments with varying building heights and densities. Each UAV covers one cluster of IoT devices, and devices forward updates to the nearest UAV. The UAVs operate in half-duplex mode and must alternate between receiving from IoT devices and transmitting to the BS. Adjacent clusters use different frequency bands, treating inter-cell interference as noise, while intra-cluster communication is orthogonal. This setting jointly optimizes UAV altitude, scheduling of updates from IoT devices to UAVs, and retransmission from UAVs to the BS. The ultimate goal is to minimise the expected weighted sum of AoI under varying channel and environmental conditions. This example is notable for being the first to analyse AoI in relay-based UAV networks under realistic transmission constraints.

This corresponds to the MDP formulation discussed in *Example 2* under *Scheduling Optimisation and Altitude Control* from the previous Chapter. The state space includes the IoT devices' AoI, virtual queue status, buffer status, achievable rate, and update size. For the actions, the UAV flies upward, downward, or maintains its altitude based on the IoT devices' updated transmit status and the virtual queue schedule.

The objective of the work in [21] is to obtain an online control policy π^* that manages the altitude h_u and UAV scheduling policy b at each time step t to minimise the Expected Weighted Sum AoI (EWSA). An online DRL approach is leveraged to solve this problem without full knowledge of channel conditions. The agent learns the control policy online using PPO, which is a stable algorithm for actor training. The authors structure the PPO algorithm into three sections (Algorithm 3.11).

The first section outlines the initialization phase, where the input and output parameters of the algorithm are defined, and the DNN parameters θ are initialized.

Algorithm 3.11 Proposed PPO Solution for Dynamic UAV Altitude Control and Scheduling Policy. Adapted from [21]. Notation and structure were standardized for this book

Require: policy network π_θ, old policy network $\pi_{\theta_{\text{old}}}$, value network V_θ; clip threshold ε; discount factor γ; Adam optimizer parameters (β_1, β_2); number of episodes L; maximum time steps $T_{\max}$; number of epochs N_{epochs}; minibatch size N_{mb}

Initialize DNN parameters θ; $\theta_{\text{old}} \leftarrow \theta$
for each training iteration **do**
  Clear episode buffer $\mathcal{B}$
  for episode $= 1$ to L **do**
    for $t = 0$ to $T_{\max} - 1$ **do**
      observe state s_t
      sample action $a_t \sim \pi_{\theta_{\text{old}}}(\cdot|s_t)$
      execute a_t (UAV altitude h_u and scheduling policy b), observe reward r_t and next state s_{t+1}
      if altitude constraint violated **then**
        Apply penalty P to r_t
      end if
      store transition (s_t, a_t, r_t, s_{t+1}) in buffer $\mathcal{B}$
    end for
  end for
  compute advantage estimates $\hat{A}_t$ (e.g., GAE-λ) and returns $\hat{G}_t$
  for epoch $= 1$ to N_{epochs} **do**
    form minibatches of size N_{mb}
    $r_\theta(s, a) = \frac{\pi_\theta(a|s)}{\pi_{\theta_{\text{old}}}(a|s)}$
    $\mathcal{L}_{\text{clip}} = \mathbb{E}[\min(r_\theta \hat{A}, \text{clip}(r_\theta, 1-\varepsilon, 1+\varepsilon)\hat{A})]$
    $\mathcal{L}_V = \mathbb{E}[(V_\theta(s) - \hat{G})^2]$, $\mathcal{L}_H = \mathbb{E}[H(\pi_\theta(\cdot|s))]$
    maximize $\mathcal{L}_{\text{PPO}} = \mathcal{L}_{\text{clip}} + c_V(-\mathcal{L}_V) + c_H\mathcal{L}_H$ via Adam
  end for
  $\theta_{\text{old}} \leftarrow \theta$; clear buffer
end for
return optimal altitude control h_u^*, scheduling policy b^*, and EWSA

The training is conducted over several iterations. In each iteration, the algorithm alternates between a data sampling phase, involving the collection of $N_{episodes}$ episodes, and an optimisation phase, where this collected data is used to perform multiple optimisation steps to update the policy parameters. Each iteration thus consists of multiple episodes, with each episode corresponding to multiple sequences of states s_t, actions a_t, and rewards r_t.

The second section details the sampling phase, where the UAV observes the AoI A_t, queue status q, and achievable rate R (between the IoT and UAV, and UAV-to-BS). Optimisation constraints relate to the time elapsed in the UAV's virtual queue and a binary variable indicating whether the selected stream from the virtual queue has a non-empty queue. Altitude constraints are respected primarily via the action space definition. The UAV selects actions a_t based on the current policy π^*, where a_t represents the UAV altitude h_u and IoTD scheduling policy b. An online algorithm further guides the agent to discourage actions that violate the altitude constraint by applying a penalty P to the overall reward r_t. This reinforces the importance of adhering to the defined altitude limits.

The UAV evaluates the reward r_t, which is defined as the negative sum of all IoTDs' AoI A_t to reflect the objective of minimising AoI. Training trajectories (s_t, a_t, r_t, s_{t+1}) are collected based on the number of episodes and rollout steps taken by each episode between updates. These data are stored for the next phase, and a message-passing interface protocol is used to achieve parallel processing among episodes. The advantage function $\hat{A}(s_t, a_t)$ is estimated for each rollout step to facilitate efficient training.

Finally, the third section details the optimisation phase, where the policy network parameters θ are updated for every epoch. The PPO clip objective function $L(\theta)$ is computed for each epoch, and the policy π_θ is optimised using stochastic gradient ascent on the stored sample data, leveraging the PPO clip objective function with learning rate α and Adam (parameters β_1, β_2). An entropy bonus is also added to the objective function to encourage exploration. After optimising the network parameters, the old policy $\pi_{\theta_{old}}$ is set to the current policy π_θ. All stored samples are then removed before the next iteration begins.

Results show that the proposed online algorithm is significantly more effective than the baseline algorithms, which involve Random Deployment of UAVs with Random Scheduling of IoT devices, and Heuristic Deployment of UAVs with Greedy Scheduling of IoT devices for uploading status update packets.

3.2.2.3 Example 3: Scheduling Optimisation and Altitude Control

This example [22] considers a dense urban environment where tall buildings obstruct direct wireless communication between the BS and IoT devices. To overcome these NLoS conditions, UAVs equipped with RIS are deployed to improve channel quality. Unlike static RIS installations, UAV-mounted RIS can dynamically adapt to mobile IoT devices in changing environments. The setup includes multiple clusters of IoT devices, each with varying sizes, where the UAV-RIS acts as a passive relay to establish virtual links between the BS and IoT clusters. A ground control station interacts with the UAV-RIS, capturing spatial and activation data of the IoT devices and managing UAV positioning, RIS phase shift control, and transmission scheduling. A generalized RIS phase shift model is used to enhance the SNR based on RIS orientation and element configuration. The main objective is to minimise the average sum AoI by jointly optimising the UAV's 3D location, RIS phase shift, and IoT device scheduling, while also ensuring secure beamforming or throughput maximization. This approach addresses the complexities of large-scale wireless networks in highly obstructed environments using DRL with structured and constrained actions.

This scenario corresponds to the MDP formulation described in *Example 3* under *Scheduling Optimisation and Altitude Control* from the previous Chapter. The state space includes each IoT device's AoI, the UAV's altitude, and the uplink transmission value. For actions, the UAV moves upward, downward, or hovers at each time step;

additionally, the cluster's activity is scheduled based on criteria such as AoI and SNR information.

Algorithm 3.12 PPO-Based Algorithm for Average-Sum-AoI Minimisation. Adapted from [22]. Notation and structure were standardized for this book

Require: Policy network π_θ, old policy network $\pi_{\theta_{\text{old}}}$, value network V_θ; PPO hyperparameters: discount γ, GAE parameter λ, clip fraction ε, learning rate α, Adam parameters (β_1, β_2), value and entropy coefficients c_V, c_H; $N_{\text{episodes}}, T_{\max}, N_{\text{epochs}}, N_{\text{mb}}$

Initialize DNN parameters θ; $\theta_{\text{old}} \leftarrow \theta$

for episode $= 1$ to N_{episodes} **do**

  Reset environment and state s_0; clear buffer $\mathcal{D}$

  for $t = 0$ to $T_{\max} - 1$ **do**

    Select action $a_t \sim \pi_{\theta_{\text{old}}}(\cdot|s_t)$; execute a_t with altitude/action validity checks

    Update AoI: $A_k(t+1) = \begin{cases} 1, & \Upsilon_k(t) \geq \Upsilon_{\min},\ I_k(t) = 1,\ W_k(t) = 1 \\ A_k(t) + 1, & \text{otherwise} \end{cases}$

    Compute reward: $r_t = -\sum_{k=1}^{K} A_k(t)$

    Observe next state s_{t+1}

    Store transition (s_t, a_t, r_t, s_{t+1}) in buffer $\mathcal{D}$

  end for

  Compute advantage estimates $\hat{A}_t$ (GAE-λ) and returns $\hat{G}_t$ from $\mathcal{D}$

  $\delta_t = r_t + \gamma V_{\theta_{\text{old}}}(s_{t+1}) - V_{\theta_{\text{old}}}(s_t)$; $\hat{A}_t = \sum_{l \geq 0} (\gamma\lambda)^l \delta_{t+l}$; $\hat{G}_t = \hat{A}_t + V_{\theta_{\text{old}}}(s_t)$

  for epoch $= 1$ to N_{epochs} **do**

    Sample minibatch from $\mathcal{D}$

    Compute ratio: $r_\theta = \frac{\pi_\theta(a|s)}{\pi_{\theta_{\text{old}}}(a|s)}$

    Compute PPO objective: $\mathcal{L}_{\text{PPO}} = \mathbb{E}[\min(r_\theta \hat{A}, \text{clip}(r_\theta, 1-\varepsilon, 1+\varepsilon)\hat{A})] + c_V(-\mathcal{L}_V) + c_H \mathcal{L}_H$

    Update parameters θ by maximizing $\mathcal{L}_{\text{PPO}}$ via Adam

  end for

  $\theta_{\text{old}} \leftarrow \theta$

end for

return policy minimizing average-sum AoI

The algorithm in [22] begins by initializing all input parameters, including the actor and critic network parameters θ, the old policy $\pi_{\theta_{\text{old}}}$ (which is set to match π_θ), and the memory buffer $\mathcal{D}$. At the start of each episode (one of $N_{episodes}$ collected for training), the UAV-RIS's (Unmanned Aerial Vehicle equipped with a Reconfigurable Intelligent Surface) initial altitude $h_u(0)$ (part of its 3D location $q_u(0)$) is reset, and the Age of Information (AoI, $A_k(0)$) for each IoTD k is also reset to one. The algorithm proceeds by defining the state s_t at time t, which includes the current AoI $A_k(t)$ of the IoTDs, their Signal-to-Noise Ratio (SNR, $\Upsilon_k(t)$), and the UAV-RIS's 3D location $q_u(t)$.

Next, the algorithm selects an action a_t from the action space $\mathcal{A}$ by observing the probabilities from the current policy $\pi_\theta(a_t|s_t)$. The action a_t comprises two components: $I(t)$ for the *switch/stay* decision within the current IoT device cluster, and $\Delta h_u(t)$ for the vertical movement (*hover*, *descend*, or *ascend*, adjusting altitude $h_u(t)$). Invalid actions, such as those exceeding altitude bounds, are explicitly eliminated.

After the action a_t is executed, the environment transitions. The SNR $\Upsilon_k(t)$ for all IoTDs is then computed. The IoTD cluster is scheduled for transmission

(b_t), and the UAV-RIS's altitude is updated to $h_u(t+1)$ based on the chosen action. The AoI $A_k(t+1)$ for each IoTD is subsequently updated. The reward r_t for this time step is computed as $r_t = -\sum_{k=1}^{K} A_k(t)$, reflecting the objective to minimise Average-Sum-AoI (ASoA). The next state s_{t+1} (comprising $A_k(t+1)$, $\Upsilon_k(t+1)$, and $q_u(t+1)$) is observed, and a done flag $done(t)$ indicates if the episode has reached its maximum duration T_{max}. The complete transition tuple $(s_t, a_t, r_t, V_\theta(s_t), \pi_\theta(a_t|s_t), s_{t+1}, done(t))$ is then stored in memory $\mathcal{D}$.

Once sufficient samples have been collected (i.e., when the step count exceeds L), the learning phase begins. Advantage estimates $\hat{A}_t$ are computed. The values recorded under the old policy $\pi_{\theta_{\text{old}}}$ are used for this computation. Minibatches of experiences are then sampled from memory $\mathcal{D}$. For each minibatch and for each epoch (up to N_{epochs}), the states s_t are passed through the actor and critic neural networks (with current parameters θ). The PPO clipped surrogate loss $L^{\text{CLIP}}(\theta)$, the critic loss $L^{\text{Critic}}(\theta)$, and the total loss $L(\theta) = L^{\text{CLIP}}(\theta) - c_1 L^{\text{Critic}}(\theta)$ are then computed as defined in the algorithm. The policy parameters θ are updated by applying stochastic gradient descent on $L(\theta)$ using the Adam optimizer with learning rate α. This optimisation process is repeated for N_{epochs} epochs to refine the network weights.

After the learning phase is complete, the old policy $\pi_{\theta_{\text{old}}}$ is updated to match the current policy π_θ, and the memory buffer $\mathcal{D}$ is cleared before the next training iteration begins. The Average-Sum-AoI (ASoA) is calculated at the end of each episode. Results show that the proposed online algorithm is significantly more effective than the baseline algorithms, which involve Random Deployment of UAVs with Random Scheduling of IoT devices, and Heuristic Deployment of UAVs with Greedy Scheduling of IoT devices for uploading status update packets. Results show the superiority of this on-Policy algorithm, PPO, as compared to DQN, an off-policy approach, with regards to stability, and speed of convergence.

3.2.3 Examples of Actor-Critic-Based Solutions and Algorithms

In this section, examples of actor-critic-based solutions to the AoI minimisation problem in UAV-assisted IoT are provided (refer to Table 3.4 for a summary of the examples). DDPG is a prominent actor-critic algorithm that adapts the successful ideas behind Deep Q-learning to continuous action domains. It is a model-free, off-policy actor-critic algorithm that utilizes deep function approximators to learn policies in high-dimensional continuous action spaces. Consequently, it can be applied to more complex problems and larger networks. DDPG is appealing because DQN cannot be directly applied to continuous action spaces. This limitation arises from the near impracticality of optimising the action at each time step in very large continuous action spaces [28].

Table 3.4 Summary of PPO-based solutions and algorithms based on different optimisation objectives

Refs.	Subsection	Objectives	States	Actions	Rewards and their constituents
[23]	3.2.3.1	Trajectory	UAV's position	UAV's flying direction, distance from SN	Negative sum of AoI
[24]	3.2.3.2	Trajectory, Scheduling	UAV's position, remaining power	UAV's flight	Average AoI, optimal IoTD energy consumption, optimal UAV propulsion energy
[25]	3.2.3.3	Trajectory, Scheduling	UAV's position, residual energy, nodes' AoI, packet lifetime, queue length	UAV's movement	Least weighted sum of: SNs' average AoI, UAV's energy consumption
[26]	3.2.3.4	Trajectory, Scheduling, RIS Phase Shift	UAV's coordinates, IoTDs' AoI	RIS controller movement	IoTDs' AoI, optimal RIS control

DDPG builds upon the Deterministic Policy Gradient (DPG) algorithm [27], leveraging the role of neural networks as function approximators to enable learning in large action spaces. It employs mini-batches to address the challenges posed by optimisation algorithms that assume samples are independently and identically distributed when using neural networks in RL. This is managed through a replay buffer, which is a finite-sized cache. The algorithm benefits from learning on a set of uncorrelated transitions, provided the replay buffer is sufficiently large. The actor and critic networks are updated at each time step by sampling a minibatch uniformly from the buffer. The target network mechanism from DQN is modified for actor-critic learning by using soft target updates instead of directly copying the weights. This is achieved by maintaining copies of the actor and critic networks for computing target values. The weights of these target networks are updated slowly to track the learned networks, thereby constraining target values to change gradually and improving learning stability. Furthermore, considering the challenge of learning from low-dimensional feature vector observations with different physical units and varying ranges across environments, learning can become difficult, and it may be hard to find hyperparameters that generalize well across different scales of state values. To address this, DDPG adopts batch normalization, where each dimension is normalized across the samples in a minibatch to ensure unit mean and variance. Batch normalization facilitates effective learning across tasks with varying units, without requiring manual normalization of inputs. To address the problem of exploration in continuous action spaces, DDPG, as an off-policy algorithm, treats exploration as separate from the learning process. Exploration is achieved by adding noise sampled from a noise process, which can be selected to suit the environment [28]. A typical representation of the DDPG algorithm is provided in Fig. 3.10.

3.2.3.1 Example 1: Trajectory Optimisation

This example [23] considers a space-air-ground integrated network comprising ground-level IoT devices, aerial UAVs, and space-based satellites. The UAVs collect sensed data from IoT devices and forward it to satellites. The focus is on optimising UAV trajectory and the matching of UAVs to IoT devices using a combination of DRL and a matching game approach. The satellites coordinate UAV-IoT pairing and facilitate trajectory planning to ensure continuous data collection and efficient information relay. The main objective is to minimise the AoI while determining the UAV's optimal flight path and destination. DRL is employed to learn the optimal trajectory that yields the minimal AoI during data collection and satellite transmission. To explore more feasible solutions, the soft actor-critic (SAC) algorithm is utilized by the UAV to learn a policy that minimizes AoI for each IoT device and determines the optimal destination. This example utilizes the MDP from *Example 3* under *Trajectory Optimisation* in the previous Chapter. The state includes the UAV's position, while the actions involve its flight direction and distance from SNs.

The algorithm aims to learn an optimal policy π_ϕ^* via maximum entropy reinforcement learning. The input consists of the initial UAV location $q_u(0)$ and IoTD locations $\{q_k\}_{k=1}^K$. The algorithm's output includes the optimal UAV trajectory q_u^* and the corresponding minimal AoI A^*. During the initialization phase, the actor network parameters ϕ, critic network parameters θ_1, θ_2, and their respective target network parameters θ_1', θ_2' are set. Additionally, the entropy coefficient α_{temp}, learning rate η, and replay buffer $\mathcal{D}$ are initialized. The target network parameters are initially set to match their corresponding online network parameters ($\theta_i' \leftarrow \theta_i$). The training proceeds for a maximum of K_{max} episodes.

For each episode k (up to K_{max}), the UAV's 2D position $q_u(0)$ is reset, initializing the state $s_0 = (x_u(0), y_u(0))$. The agent then interacts with the environment in a series of time steps. At each time step t, the agent samples an action $a_t = (d_t, \theta_t)$ from the current policy $\pi_\phi(\cdot|s_t)$, where d_t is the flight distance and θ_t is the flight direction. Executing this action moves the UAV, leading to a new state s_{t+1}. An AoI-based reward $r_t = -A_m(t)$ is computed, where $A_m(t)$ represents the minimum AoI achieved at that time step for the relevant IoTDs. The transition tuple (s_t, a_t, r_t, s_{t+1}) is then stored in the replay buffer $\mathcal{D}$.

Periodically, or after each environmental step, a minibatch of transitions $\{(s_j, a_j, r_j, s_{j+1})\}$ is uniformly sampled from the replay buffer $\mathcal{D}$ to update the networks. The algorithm constructs two Q-function critics, Q_{θ_1} and Q_{θ_2}, and their corresponding target networks, $Q_{\theta_1'}$ and $Q_{\theta_2'}$. For each sampled transition in the minibatch,

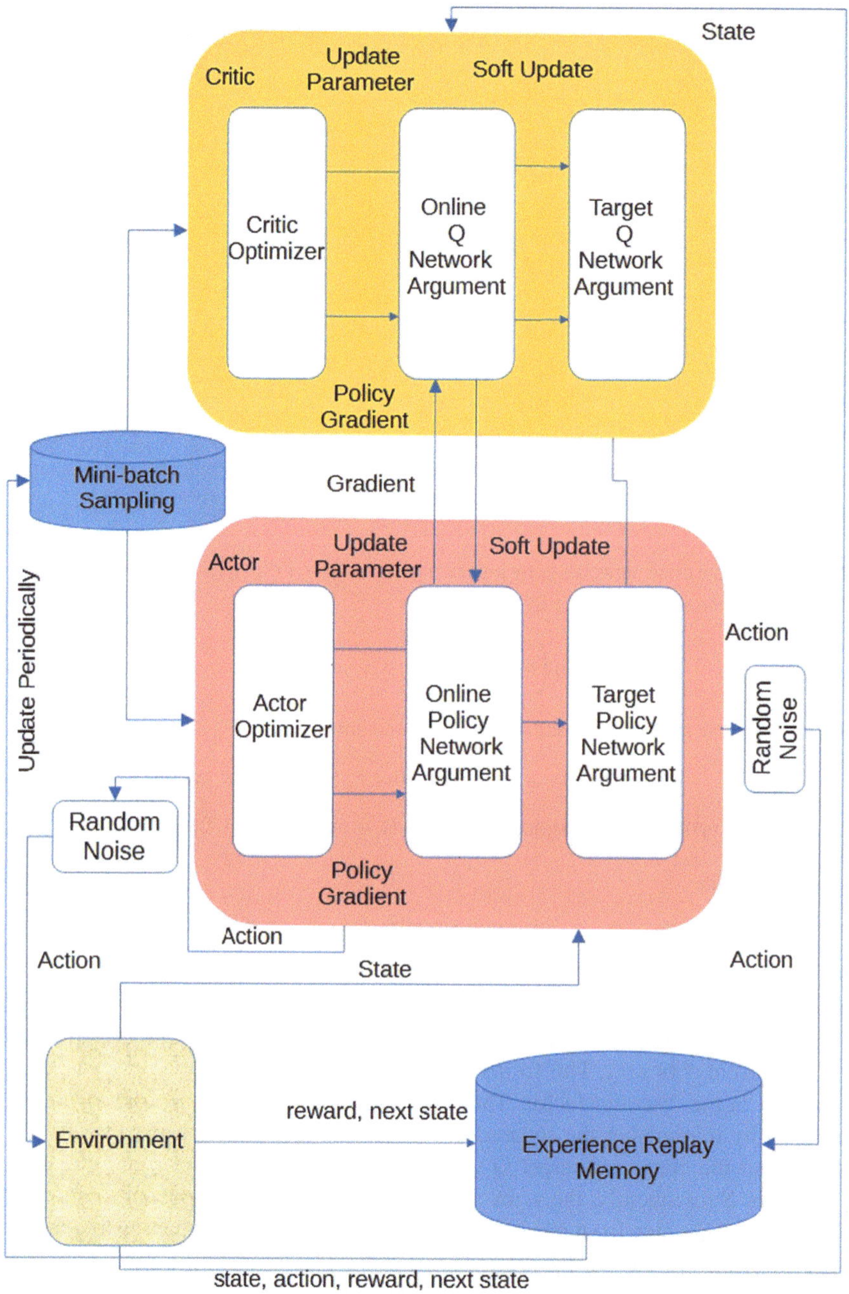

Fig. 3.10 Typical representation of the deep deterministic policy gradient (DDPG) architecture

Algorithm 3.13 SAC-Based AoI Optimisation Algorithm. Adapted from [23]. Notation and structure were standardized for this book

Require: Initial UAV location $q_u(0)$, IoTD locations $\{q_k\}_{k=1}^K$
Ensure: Optimal UAV trajectory q_u^* and corresponding minimal AoI A^*
Initialize actor network π_ϕ, critic networks $Q_{\theta_1}, Q_{\theta_2}$, target critics $Q_{\theta_1'}, Q_{\theta_2'}$
Initialize entropy coefficient α_{temp}, learning rate η, replay buffer $\mathcal{D}$
Set target parameters: $\theta_i' \leftarrow \theta_i$, for $i \in \{1, 2\}$
for episode $k = 1$ to $K_{\max}$ **do**
  Reset UAV position $q_u(0)$, initialize state $s_0 = (x_u(0), y_u(0))$
  while not terminal **do**
    Sample action $a_t = (d_t, \theta_t) \sim \pi_\phi(\cdot|s_t)$
    Execute a_t, observe new state s_{t+1} and AoI-based reward $r_t = -A_m(t)$
    Store transition (s_t, a_t, r_t, s_{t+1}) in buffer $\mathcal{D}$
    Sample minibatch $\{(s_j, a_j, r_j, s_{j+1})\}$ from buffer $\mathcal{D}$
    Sample next action $\tilde{a}_{j+1} \sim \pi_\phi(\cdot|s_{j+1})$ and calculate log probability $\log \pi_\phi(\tilde{a}_{j+1}|s_{j+1})$
    Compute target Q-value:

$$\hat{Q}(s_j, a_j) = r_j + \gamma \left(\min_{i=1,2} Q_{\theta_i'}(s_{j+1}, \tilde{a}_{j+1}) - \alpha_{temp} \log \pi_\phi(\tilde{a}_{j+1}|s_{j+1}) \right)$$

    Update critic networks θ_i by minimizing $J_{\theta_i} = \frac{1}{2}\mathbb{E}_{\mathcal{D}} \left(Q_{\theta_i}(s_j, a_j) - \hat{Q}(s_j, a_j) \right)^2$
    Update actor network ϕ by minimizing

$$J_\pi = \mathbb{E}_{\mathcal{D}} \left(\alpha_{temp} \log \pi_\phi(a_j|s_j) - \min_{i=1,2} Q_{\theta_i}(s_j, a_j) \right)$$

    Update temperature coefficient α_{temp} by minimizing

$$J(\alpha_{temp}) = \mathbb{E}_{a\sim\pi_\phi(\cdot|s)} \left[-\alpha_{temp}(\log \pi_\phi(a|s) + \text{target\_entropy}) \right]$$

    Soft update target networks: $\theta_i' \leftarrow \tau\theta_i + (1 - \tau)\theta_i', \quad \forall i \in \{1, 2\}$
    $s_t \leftarrow s_{t+1}$
  end while
end for

a target value $V_{\bar{\theta}}(s_{j+1})$ is estimated by evaluating the expected minimum Q-value from the target critic networks and subtracting the entropy regularization term based on the current policy π_ϕ. This forms the target Q-value $\hat{Q}(s_j, a_j) = r_j + \gamma V_{\bar{\theta}}(s_{j+1})$, where γ is the discount factor. The critic network parameters θ_1 and θ_2 are then updated by minimising the squared difference between their current Q-value estimates and the computed target Q-values using the learning rate η and the Adam optimizer. Subsequently, the actor network parameters ϕ are updated to maximise the expected Q-value from one of the critic networks while also maximising the policy's entropy, again using the learning rate η and the Adam optimizer. The entropy coefficient α_{temp} is automatically adjusted during training to ensure the policy maintains a predefined target entropy H_0. This is achieved by minimising a specific loss function for α_{temp}. Finally, the target critic network parameters θ_i' are slowly updated using a soft update rule: $\theta_i' \leftarrow \tau\theta_i + (1 - \tau)\theta_i'$, where τ is a small update rate, ensuring stability in learning.

After the entire training process is complete, the optimal UAV trajectory q_u^* and the corresponding minimal AoI A^* are obtained. In the broader system, the satellite can then execute pairing between the UAV and IoTD using the Gale-Shapley algorithm, leveraging the learned optimal trajectory.

Results show a significant reduction in AoI compared to DQN, as well as a reduction in AoI with the deployment of more UAVs.

3.2.3.2 Example 2: Trajectory and Scheduling Optimisation

This example [24] considers an AoI and energy-aware UAV-assisted data collection framework where a group of IoT devices is unable to directly transmit data to a remote base station due to energy constraints and poor channel conditions. A UAV, operating in half-duplex mode, is deployed as a mobile data collector. The UAV alternates between hovering and traveling phases in each Hovering-Traveling Process (HTP) to gather data. The system aims to jointly minimise the expected weighted sum of the average AoI, UAV propulsion energy, and IoT device transmission energy. optimisation variables include the UAV's flight speed, hovering locations, and bandwidth allocation. Given the dynamics and constraints of the system, the problem is modeled as an MDP. This example utilizes the MDP in *Example 4* under *Trajectory and Scheduling Optimisation* in the previous Chapter. The state space is characterised by the UAV's location, the current AoI of all IoTDs, and the coverage condition indicator, which indicates whether an IoTD has been served in the hovering-traveling process (HTP) of the current flight round. One HTP includes a hovering duration and a flying duration. The action space includes UAV flight distance, flight direction, and UAV speed.

DDPG suffers from overestimated Q-values, leading to invalid learned policies. To address this, Twin Delayed DDPG (TD3) was proposed in [24], effectively handling multidimensional variables and providing a solution to Q-value overestimation, ensuring more accurate learned policies with actor network μ (parameterized by θ_μ) and critic networks Q_1, Q_2 (parameterized by $\theta_{Q_1}, \theta_{Q_2}$).

The algorithm (Algorithm 3.14) works as follows: Initially, the discount factor γ, soft update rate υ, penalty $h^{penalty}$, and award h^{award} are initialized. The online actor network parameters θ_μ and two online critic network parameters $\theta_{Q_1}, \theta_{Q_2}$ are initialized, along with their corresponding target network parameters $\theta'_\mu, \theta'_{Q_1}, \theta'_{Q_2}$. The algorithm utilizes a replay buffer $\mathcal{D}$ to store experiences (s_t, a_t, r_t, s_{t+1}). The mini-batch size N_{batch}, maximum number of episodes $N_{episodes}$, and maximum steps per episode T_{max} are also configured.

For each episode (up to $N_{episodes}$), the environment is reset: the UAV's initial location $q_u(0)$, the AoI $A_k(0)$ of all IoTDs, and their coverage flags $C_k(0)$ are initialized, forming the initial state s_0. For each timestep t (from 0 to T_{max}), the UAV agent selects an action a_t by adding clipped random noise $\mathcal{N}_t$ to the output of its current policy $\mu(s_t|\theta_\mu)$. The UAV executes the selected action a_t, leading to a new state s_{t+1}, and receives an associated reward r_t. The transition tuple (s_t, a_t, r_t, s_{t+1}) is then stored in the replay buffer $\mathcal{D}$.

Algorithm 3.14 TD3-based AoI-Energy-Aware UAV Trajectory Planning Algorithm. Adapted from [24]. Notation and structure were standardized for this book

Require: Discount factor γ, soft update rate υ, mini-batch size N_{batch}, max episodes $N_{episodes}$, max steps T_{max}
Ensure: Optimal UAV trajectory q_u^* and deterministic policy $\mu(s_t)$

```
Initialize online actor and critic networks: θ_μ, θ_Q1, θ_Q2
Initialize target actor and critic networks: θ'_μ, θ'_Q1, θ'_Q2
Initialize replay memory D
for episode = 1 to N_episodes do
    Reset environment and initialize state s_0
    for timestep t = 0 to T_max do
        Select action a_t = μ(s_t|θ_μ) + clip(N(0, σ), −c, c)
        Execute a_t, observe reward r_t and next state s_{t+1}
        Store transition (s_t, a_t, r_t, s_{t+1}) in D
        Sample mini-batch {(s_i, a_i, r_i, s_{i+1})}_{i=1}^{N_batch} from D
        Add clipped noise: ã_i = μ(s_{i+1}|θ'_μ) + clip(N(0, σ), −c, c)
        Compute target value y_i = r_i + γ min_{j=1,2} Q_j(s_{i+1}, ã_i|θ'_{Q_j})
        Update critic networks by minimizing L_{Q_j} = (1/N_batch) Σ_{i=1}^{N_batch} (Q_j(s_i, a_i|θ_{Q_j}) − y_i)^2
        if t mod Z = 0 then
            Update actor policy via deterministic policy gradient
            Soft-update target networks: θ'_X ← υθ_X + (1 − υ)θ'_X
        end if
        Set s_t ← s_{t+1}
    end for
end for
```

Subsequently, a mini-batch of N_{batch} transitions $\{(s_i, a_i, r_i, s_{i+1})\}$ is uniformly sampled from the memory buffer $\mathcal{D}$. To compute the target Q-values, a noisy target action $\widetilde{a}_i$ is derived by adding clipped noise to the output of the target actor network, i.e., $\widetilde{a}_i = \mu(s_{i+1}|\theta'_\mu) + \text{clip}(\mathcal{N}(0, \sigma), -c, c)$. The target value y_i for each transition is then computed as $y_i = r_i + \gamma \min_{j=1,2} Q_j(s_{i+1}, \widetilde{a}_i|\theta'_{Q_j})$. The parameters of the online critic networks $\theta_{Q_1}, \theta_{Q_2}$ are updated by minimising their respective loss functions L_Q, which are the squared differences between their current Q-value estimates and the computed target values y_i.

The online actor network parameters θ_μ are updated using the deterministic policy gradient method to maximise the estimated Q-value from the first critic network. Crucially, the target network parameters $\theta'_\mu, \theta'_{Q_1}$, and θ'_{Q_2} are updated less frequently, specifically every Z steps, using a soft update rule: $\theta'_X \leftarrow \upsilon\theta_X + (1 - \upsilon)\theta'_X$, where υ is the soft update rate and X denotes either μ, Q_1, or Q_2. This delayed update, along with target policy smoothing, helps to prevent Q-value overestimation. After the entire training process is complete, the optimal UAV trajectory q_u^* and the corresponding deterministic policy $\mu(s_t)$ are obtained.

The results show that the proposed algorithm outperforms DQN and A2C-based [29] algorithms in terms of both achievable AoI and energy efficiency.

3.2.3.3 Example 3: Trajectory and Scheduling Optimisation

This example [25] introduces energy harvesting UAVs used as mobile relays to assist wireless sensor networks. It is the first to incorporate renewable energy harvesting into the UAV trajectory and scheduling optimisation framework. A rotary-wing UAV with a fixed-capacity battery switches between two modes: a working mode for data collection and a waiting mode for energy harvesting. In the working mode, the UAV flies to SNs to collect data; in the waiting mode, it descends and harvests energy until resuming operation. The UAV's flight trajectory, SN scheduling, and energy-aware operation are jointly optimized to minimise the average AoI while reducing energy consumption. The system is modeled as an MDP with different action spaces for each mode, and the objective is to derive an optimal policy that balances data freshness with energy efficiency. This example utilizes the MDP from *Example 5* under *Trajectory and Scheduling Optimisation* in the previous Chapter. The states include the UAV's position and residual energy, the SN's AoI, its latest packet lifetime, and its queue length. The actions involve the UAV's horizontal and vertical flight displacement on the grid cells, whether in waiting or working mode.

Algorithm 3.15 A3C-based UAV-Assisted Data Collection Algorithm. Adapted from [25]. Notation and structure were standardized for this book

Require: Workers w_j $(j = 1, \ldots, J)$, global actor parameters θ_μ, global critic parameters θ_V, energy threshold E_{th}, discount factor γ, learning rates α_μ, α_V, max steps T_{max}, entropy coefficient β

Ensure: Optimal trajectory q_u^*, expected cost C_π, harvested energy $E_h(t)$

for each worker w_j in parallel **do**
  Initialize local networks: $\theta_\mu^j \leftarrow \theta_\mu, \theta_V^j \leftarrow \theta_V$
  Initialize state s_0, UAV location $q_u(0)$, residual energy $E(0)$
  for timestep $t = 0$ to T_{max} **do**
    if $E(t) > E_{th}$ **then**
      Select action $a_t \sim \pi(a_t|s_t; \theta_\mu^j)$ and execute
      Observe reward r_t and next state s_{t+1}
    else
      UAV hovers to harvest energy
      Update AoI, cost, and residual energy. Observe reward r_t and next state s_{t+1}
    end if
    Compute advantage: $A_t = r_t + \gamma V(s_{t+1}; \theta_V^j) - V(s_t; \theta_V^j)$
    Update local actor: $\theta_\mu^j \leftarrow \theta_\mu^j + \alpha_\mu \nabla_{\theta_\mu^j} \log \pi(a_t|s_t) A_t + \beta \nabla_{\theta_\mu^j} H(\pi(s_t))$
    Update local critic: $\theta_V^j \leftarrow \theta_V^j - \alpha_V \nabla_{\theta_V^j} (A_t)^2$
    if local update step mod $K = 0$ **then**
      Push local gradients to global networks and update local networks with new global parameters
    end if
    $s_t \leftarrow s_{t+1}$
  end for
end for
return Optimal trajectory q_u^*, expected cost C_π, and harvested energy $E_h(t)$

The authors consider an energy-harvesting UAV-assisted WSN, employing advanced DRL methods due to the large state space and unknown transition probabilities of the environment. To find the optimal policy π^*, DQN uses two neural networks with the same architecture (target and evaluation with parameters θ'_Q and θ_Q). The parameters of the evaluation network θ_Q are updated in real time, while the target network parameters θ'_Q are updated regularly to ensure network stability.

The Asynchronous Advantage Actor-Critic (A3C), a multithreaded version of actor-critic, involves an actor (with parameters θ_μ) that selects actions a_t at each time step t, constantly adjusting and optimising the policy π, and a critic (with parameters θ_V) that evaluates the value $V(s_t)$ of the action at a state s_t.

If the residual energy $E(t)$ of the UAV is less than or equal to a set energy threshold E_{th}, the UAV position $q_u(t)$ and energy $E(t)$ are updated after an action a_t (horizontal/vertical flight or choice of uploading node, depending on the mode of operation) is taken based on the policy π. Otherwise, the UAV remains in the air, harvests energy $E_h(t)$, refreshes itself, updates the AoI $A(t)$, calculates the cost $C(t)$, and proceeds to the next state s_{t+1}. Exploratory actions are increased via entropy $H(\pi)$. The learning rates for the actor and critic are α_μ and α_V, respectively.

Results indicate that the A3C algorithm, when implemented with eight parallel threads, achieves convergence up to 60 times faster than the traditional DQN and exhibits enhanced adaptability to diverse energy harvesting scenarios.

3.2.3.4 Example 4: Trajectory, Scheduling and RIS Phase Shift Optimisation

This example [26] addresses UAV-assisted data collection in a realistic urban environment with building-induced signal blockage. A rotary-wing UAV collects status updates from IoT devices using TDMA to prevent interference, and a RIS is deployed on high-rise buildings to improve channel quality. Unlike traditional approaches that use probabilistic LoS models, this work uses urban environment characteristics to determine link connectivity. The problem is modeled as an MDP, where the state captures UAV location and IoT device AoI values, the action includes UAV speed, direction, and IoTD scheduling, and the reward measures the AoI reduction with penalties for invalid UAV moves. The RIS controller adjusts the phase shift dynamically to enhance UAV-IoTD links. The objective is to jointly optimize UAV trajectory, IoTD scheduling, and RIS configuration to reduce overall network AoI. This example utilizes the MDP discussed under *Trajectory, Scheduling and RIS Phase Shift Optimisation* in the previous Chapter. The states include the UAV's coordinates and the AoI of IoTDs, while the actions involve the RIS controller's movement (based on the UAV's average flying speed and horizontal direction) and the IoTDs' transmission schedules.

Algorithm 3.16 SAC-AO-RIS Algorithm. Adapted from [26]. Notation and structure were standardized for this book

Require: Actor network Φ, critic networks ϕ_1, ϕ_2, value network V, target value network V', replay buffer $\mathcal{D}$
Ensure: RIS phase shift control ϑ^*, UAV trajectory q_u^*, IoTD scheduling policy b^*
for each episode = 1 to N_{episodes} **do**
  Initialize environment; observe initial state s_0
  for each time step $t = 1$ to T **do**
    Observe current state s_t
    Sample action $a_t \sim \pi_\Phi(a_t|s_t)$
    Determine RIS phase shift ϑ_t using AO method
    Execute UAV movement and IoTD scheduling policy b_t
    Observe new state s_{t+1} and reward r_t
    if UAV violates boundary constraints **then**
      Apply penalty P, cancel movement, update s_{t+1} and r_t
    end if
    Store (s_t, a_t, r_t, s_{t+1}) in replay buffer $\mathcal{D}$
    if Terminal state reached **then**
      Sample a batch from $\mathcal{D}$ using Prioritized Experience Replay
      Update networks: value network V, critic networks ϕ_1, ϕ_2, and actor network Φ
      Soft update target value network: $V' \leftarrow \tau V + (1 - \tau)V'$
    end if
  end for
end for

In [26], the authors aim to minimise the AoI of IoTDs via joint optimisation of UAV trajectory q_u, IoTD scheduling b, and discrete RIS phase shift ϑ based on Algorithm 3.16. To solve this problem, the authors initialised DNN parameters θ and parameters for actor network Φ, critic networks ϕ_1, ϕ_2, value network V, and target value networks V', and experience replay buffer $\mathcal{D}$.

For each episode, the environment is initialized, and the initial state s_0 is observed by the UAV. For each time step t, the state s_t is observed, and the action $a_t \sim \pi_\Phi(\cdot|s_t)$ is taken. The RIS phase shift ϑ_t is determined using the alternating optimisation (AO) algorithm. The UAV then executes trajectory control and IoTD scheduling policy b_t. The new state s_{t+1} is observed, and the reward r_t is computed by the UAV. A penalty P is enacted if the UAV flies out of bounds by canceling the movement, updating the next state s_{t+1}, and the reward r_t.

During the training phase, the experience (s_t, a_t, r_t, s_{t+1}) is stored in the replay buffer $\mathcal{D}$. If the terminal state is reached, the Recent Prioritized Experience Replay (RPER) technique is used to sample a batch of experiences, and the value network V, the parameters of the two soft Q-networks (ϕ_1, ϕ_2), the policy network Φ, and the target network V' are updated using learning rate α and Adam optimizer. These steps are repeated until all time steps are complete and the final episode is reached.

Extensive simulations using the proposed algorithm demonstrate a significant reduction in average episodic AoI compared to several baseline schemes. One such baseline involves using the SAC algorithm to jointly optimize the UAV's flight trajectory and the transmission scheduling of IoTDs, while the phase shifts of the RIS are randomly configured. A similar version of this setup is also considered without the use of RIS. Another baseline [10] employs a learning-based approach for optimising both the UAV trajectory and IoTD scheduling, while the RIS phase shifts are determined using the proposed AO algorithm. This scenario is also evaluated under conditions with a randomly deployed RIS, as well as without RIS entirely.

In addition to these, two fixed-trajectory baselines without RIS are examined. In the first, the UAV flies along a predetermined rectangular strip track [30] at a constant velocity and schedules the closest IoTD based on distance, a greedy scheduling approach. In the second, the UAV follows the same fixed trajectory but selects IoTDs randomly for data uploads.

Across all these scenarios, the proposed algorithm consistently outperforms the baselines in reducing the AoI, showcasing its robustness and adaptability in various network configurations and scheduling strategies.

3.3 Summary

This chapter covers the RL and DRL algorithms in general, as well as those implemented for AoI-aware applications in UAV-assisted WSN/IoT networks. The different broad classes of RL, namely value-based, policy-based, and actor-critic-based, have been described. In addition to the default single-agent RL, a multi-agent RL concept has also been introduced, whereby agents perform actions independently and collaborate. This multi-agent RL approach is deployed in both centralised and decentralised UAV/IoT data collection systems. Finally, the solutions and DRL algorithms of the MDP problems previously discussed in Chap. 2 are presented, with each solution categorised under either value, policy, or actor-critic RL class.

References

1. R.S. Sutton, A.G. Barto, *Reinforcement Learning: An Introduction*, 2nd edn. (MIT Press, 2018)
2. V. Mnih, K. Kavukcuoglu, D. Silver, A.A. Rusu, J. Veness, M.G. Bellemare, A. Graves, M. Riedmiller, A.K. Fidjeland, G. Ostrovski et al., Human-level control through deep reinforcement learning. Nature **518**(7540), 529–533 (2015)
3. J. Schulman, F. Wolski, P. Dhariwal, A. Radford, O. Klimov, Proximal policy optimization algorithms (2017), arXiv:1707.06347
4. R.J. Williams, Simple statistical gradient-following algorithms for connectionist reinforcement learning. Mach. Learn. **8**(3–4), 229–256 (1992). https://doi.org/10.1007/BF00992696
5. V. Konda, J. Tsitsiklis, Actor-critic algorithms, in *Advances in Neural Information Processing Systems*, ed. by S. Solla, T. Leen, K. Müller, vol. 12 (MIT Press, 1999)
6. I. Grondman, L. Busoniu, G. Lopes, R. Babuska, A survey of actor-critic reinforcement learning: Standard and natural policy gradients. IEEE Trans. Syst. Man Cybern. Part C (Appl. Rev.) **42**(6), 1291–1307 (2012)
7. V. Konda, Actor-critic algorithms. Ph.D. dissertation, Department of Electrical Engineering and Computer Science, Massachusetts Institute of Technology, 2002

8. O.A. Amodu, C. Jarray, R. Mahmood, H. Althumali, U.A. Bukar, R. Nordin, N.F. Abdullah, N.C. Luong, Deep reinforcement learning for AoI minimization in UAV-aided data collection for WSN and IoT: a survey. IEEE Access **12**, 108000–108040 (2024). https://doi.org/10.1109/ACCESS.2024.3425497
9. H. Van Hasselt, A. Guez, D. Silver, Deep reinforcement learning with double q-learning, in *Proceedings of the AAAI Conference on Artificial Intelligence*, vol. 30, no. 1 (2016)
10. C. Zhou, H. He, P. Yang, F. Lyu, W. Wu, N. Cheng, X. Shen, Deep RL-based trajectory planning for AOI minimization in UAV-assisted IoT, in *2019 11th International Conference on Wireless Communications and Signal Processing (WCSP)*, (IEEE, 2019), pp. 1–6
11. Z. Li, P. Tong, J. Liu, X. Wang, L. Xie, H. Dai, Learning-based data gathering for information freshness in UAV-assisted IoT networks. IEEE Internet Things J. **10**(3), 2557–2573 (2023). https://doi.org/10.1109/JIOT.2022.3215521.
12. X. Li, J. Li, D. Liu, Energy-efficient UAV trajectory design with information freshness constraint via deep reinforcement learning. Mob. Inf. Syst. **2021**, 1–9 (2021)
13. S.F. Abedin, M.S. Munir, N.H. Tran, Z. Han, C.S. Hong, Data freshness and energy-efficient UAV navigation optimization: a deep reinforcement learning approach. IEEE Trans. Intell. Transp. Syst. **22**(9), 5994–6006 (2020)
14. Q. Dang, Q. Cui, Z. Gong, X. Zhang, X. Huang, X. Tao, AOI oriented UAV trajectory planning in wireless powered IoT networks, in *2022 IEEE Wireless Communications and Networking Conference (WCNC)*, (IEEE, 2022), pp. 884–889
15. J. Xu, X. Jia, Z. Hao, Research on information freshness of UAV-assisted IoT networks based on DDQN, in *2022 IEEE 5th Advanced Information Management, Communicates, Electronic and Automation Control Conference (IMCEC)*, vol. 5 (IEEE, 2022), pp. 427–433
16. A. Ferdowsi, M.A. Abd-Elmagid, W. Saad, H.S. Dhillon, Neural combinatorial deep reinforcement learning for age-optimal joint trajectory and scheduling design in UAV-assisted networks. IEEE J. Sel. Areas Commun. **39**(5), 1250–1265 (2021)
17. P. Tong, J. Liu, X. Wang, B. Bai, H. Dai, Deep reinforcement learning for efficient data collection in UAV-aided internet of things, in *2020 IEEE International Conference on Communications Workshops (ICC Workshops)*, (IEEE, 2020), pp. 1–6
18. C. Joo, A. Eryilmaz, Wireless scheduling for information freshness and synchrony: drift-based design and heavy-traffic analysis. IEEE/ACM Trans. Netw. **26**(6), 2556–2568 (2018)
19. Z. Li, J. Liu, L. Xie, X. Wang, M. Jin, A deep reinforcement learning approach for multi-UAV-assisted data collection in wireless powered IoT networks, in *2022 14th International Conference on Wireless Communications and Signal Processing (WCSP)*, (IEEE, 2022), pp. 44–49
20. M. Samir, M. Elhattab, C. Assi, S. Sharafeddine, A. Ghrayeb, Optimizing age of information through aerial reconfigurable intelligent surfaces: a deep reinforcement learning approach. IEEE Trans. Veh. Technol. **70**(4), 3978–3983 (2021)
21. M. Samir, C. Assi, S. Sharafeddine, A. Ghrayeb, Online altitude control and scheduling policy for minimizing AoI in UAV-assisted IoT wireless networks. IEEE Trans. Mob. Comput. **21**(7), 2493–2505 (2022). https://doi.org/10.1109/TMC.2020.3042925.
22. M. Sherman, S. Shao, X. Sun, J. Zheng, Optimizing AoI in UAV-RIS-assisted IoT networks: off policy versus on policy. IEEE Internet Things J. **10**(14), 12401–12415 (2023)
23. X. Wei, G. Zhang, Z. Han, Satellite-controlled UAV-assisted IoT information collection with deep reinforcement learning and device matching, in *2022 7th International Conference on Intelligent Computing and Signal Processing (ICSP)*, (IEEE, 2022), pp. 1254–1259
24. M. Sun, X. Xu, X. Qin, P. Zhang, AOI-energy-aware UAV-assisted data collection for IoT networks: a deep reinforcement learning method. IEEE Internet Things J. (2021)
25. N. Zhang, J. Liu, L. Xie, P. Tong, A deep reinforcement learning approach to energy-harvesting UAV-aided data collection, in *2020 International Conference on Wireless Communications and Signal Processing (WCSP)*, (IEEE, 2020), pp. 93–98
26. X. Fan, M. Liu, Y. Chen, S. Sun, Z. Li, X. Guo, RIS-assisted UAV for fresh data collection in 3D urban environments: a deep reinforcement learning approach. IEEE Trans. Veh. Technol. **72**(1), 632–647 (2022)

27. D. Silver, G. Lever, N. Heess, T. Degris, D. Wierstra, M. Riedmiller, Deterministic policy gradient algorithms, in *Proceedings of the International Conference on Machine Learning (ICML)*, (PMLR, 2014), pp. 387–395
28. T.P. Lillicrap, J.J. Hunt, A. Pritzel, N. Heess, T. Erez, Y. Tassa, D. Silver, D. Wierstra, Continuous control with deep reinforcement learning (2015), arXiv:1509.02971
29. J. Hu, H. Zhang, L. Song, R. Schober, H.V. Poor, Cooperative internet of UAVs: distributed trajectory design by multi-agent deep reinforcement learning. IEEE Trans. Commun. **68**(11), 6807–6821 (2020)
30. Y. Wang et al., Trajectory design for UAV-based internet-of-things data collection: a deep reinforcement learning approach. IEEE Internet Things J. **9**(5), 3899–3912 (2022)

Chapter 4
Challenges and Future Considerations

4.1 DRL-Related Challenges in Age of Information (AoI)-Aware UAV-IoT

Although Deep Reinforcement Learning (DRL) has been successfully applied to several practical applications in recent years, particularly in UAV-assisted data collection for IoT, it still faces a number of challenges. Some of these challenges, including those discussed in [1, 2], are presented below.

4.1.1 Convergence

A DRL algorithm is considered to have converged when its learning curve flattens [3]. During this phase, the change in the loss function between iterations drops below a predefined threshold [4]. However, DRL often suffers from slow convergence and high computational costs, particularly in complex environments. In the context of UAV-assisted IoT, such an environment may involve factors like the channel characteristics, the presence of obstacles, network size, the mobility of network elements, the integration of technologies such as reconfigurable intelligent surfaces, UAV swarm coordination, unknown traffic patterns, energy harvesting, and optimisation of data transfer or scheduling.

Another challenge is overfitting and insufficient exploration. Off-policy algorithms, for instance, face longer convergence times due to methods like epsilon-greedy and the variability of Q-values [5]. Without sufficiently diverse exploration in high-dimensional spaces, the algorithm is more likely to converge to local optima. Furthermore, unstable training can hinder the convergence process, requiring careful selection of the appropriate algorithm. For example, DQN, despite utilizing neural networks to approximate the Q-function, converges slowly [6]. Therefore,

O. A. Amodu et al., *Markov Decision Processes and Reinforcement Learning for Timely UAV-IoT Data Collection Applications*, Studies in Computational Intelligence 1220,
https://doi.org/10.1007/978-3-031-97011-5_4

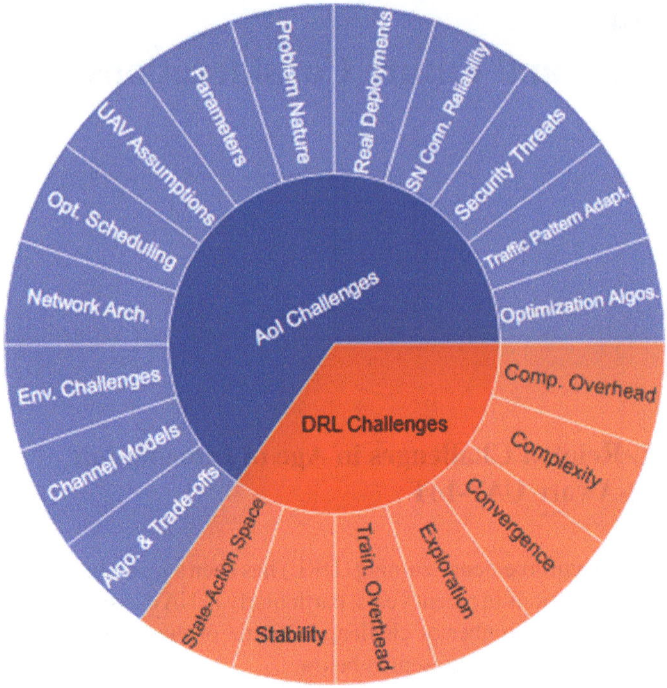

Fig. 4.1 Visual summary of common challenges in DRL and AoI in UAV-assisted IoT

future research is needed to develop more efficient algorithms that can achieve faster convergence, even in complex environments.

4.1.2 Complexity

DRL's ability to master specific action sequences may lead to poor performance in unexpected circumstances, as it can result in overfitting. Enhancing performance often entails trade-offs, such as increased complexity. The opaque nature of DRL models hinders interpretation and makes it difficult to troubleshoot problems when they arise. Modeling real-world scenarios presents challenges due to environmental complexity and stochasticity. Designing effective reward functions for multi-objective, nonlinear problems is also challenging, particularly when balancing weights and trade-offs. Exploration further makes this more complex with the combinatorial nature of hybrid discrete-continuous actions. Exploring techniques like quantum computing might be promising, given DRL's poor scalability with problem size and complexity.

4.1.3 Computation Overhead

The widespread adoption of DRL is made possible by the affordability of modern computational resources. However, UAVs, with their limited memory and energy capacities, often have low computational power, making DRL applications challenging. Optimisation and AI-based learning require substantial computational resources, which can strain UAV capabilities. Training neural networks, in particular, is computationally intensive and typically demands powerful hardware, such as expensive GPUs. This trade-off between computational power and cost is evident in multi-agent DRL schemes as well. As such, addressing these computational limitations is paramount for realizing the full potential of DRL in UAV applications.

Moreover, many IoT deployments use distributed architectures for decision-making, particularly when a central authority is absent or connectivity to centralised systems is poor. High communication latency to these systems can also render a centralised architecture challenging. This is where RL is particularly useful, as it excels at decentralised decision-making. However, RL algorithms still suffer from high computational overhead, which scales with the number of agents involved. To address this, parallel RL algorithms, which run on multiple agents in distributed environments, offer a promising solution. These algorithms, however, come with their own set of challenges, including high communication overhead and increased execution time. To mitigate these issues, novel solutions are required. For example, the partitioning algorithm proposed in [7] minimises communication overhead in parallel RL for IoT environments. This approach combines dynamic state partitioning and leverages the exploration capabilities of agents to build partition knowledge, efficiently mapping agents to partitions to reduce communication between them.

4.1.4 Training Overhead

Although acquiring sufficient training data in dynamic environments is difficult, it is essential for optimising neural networks [8]. Techniques that enhance training stability and convergence, such as experience replay and random sampling, are therefore invaluable. However, the time and computational requirements for DRL model training are often substantial, involving millions of iterations for meaningful results due to their deep learning components [9] and reliance on backpropagation. Independent training data enhances stability and convergence, making experience replay and random sampling invaluable techniques. The time and computational requirements for DRL model training are often substantial, involving millions of iterations for meaningful results due to deep learning components [9] and backpropagation. Techniques like imitation learning, which leverage heuristic algorithms, can shorten training times [10]. Both insufficient training and overtraining are undesirable [11], and hyperparameter tuning is crucial. Ensuring tractable training, particularly by reducing the action space without compromising accuracy, is vital. Training complexity escalates with multi-agent systems [12], emphasizing the need for efficient training methods and adaptability to environmental dynamics [13]. Poorly designed reward functions or action spaces can lead to training failures or excessively long learning periods. Therefore, careful design of the reward function and action space is paramount for successful and efficient training in UAV-assisted IoT applications.

4.1.5 State-Action Space Dimension

High-dimensional state-action spaces pose significant challenges for DRL. Many methods struggle with large and complex action spaces, often limited to small, discrete, or low-dimensional continuous actions. Many real-world scenarios, however, involve extensive discrete or hybrid action spaces, complicating the design of states, actions, and rewards. DRL's susceptibility to local optima in high-dimensional spaces is another challenge, particularly in navigation tasks where policy-based solutions are more suitable. For instance, DQN outputs discrete actions, making it less effective for very large action spaces, thus affecting exploration and sample efficiency.

When transition probabilities are unknown in models with vast state-action spaces, traditional MDP solutions become challenging. In large-scale scenarios, DQN struggles with optimal scheduling, necessitating methods like LSTM-based autoencoders for state space mapping. RL also finds it hard to learn with limited training data, often requiring millions of interactions for simple problems, restricting its industrial applicability. Efficient environment exploration and information collection are crucial for optimal policy learning in complex, sparse-reward environments, where various factors could interfere with the agent's ability to identify the most optimal actions.

Control policy formulation for closely related design components, like UAV altitude and RIS phase shift matrices, is particularly challenging [15]. Adding trajectory and scheduling to problems like this even makes it more challenging [14]. Such problems, often hybrid discrete-continuous action spaces, demand efficient DRL algorithms to manage the extensive possible actions, making learning more difficult.

4.1.6 Exploration

Exploration is a critical aspect of DRL, but it can be time-consuming and challenging, especially in complex environments with large state-action spaces. Approaches like epsilon-greedy, Boltzmann exploration, Upper Confidence Bound (UCB), and entropy regularisation offer various methods to balance exploration and exploitation [16, 17]. Safe and coordinated exploration in multi-agent systems, particularly under partial observation and non-stationary conditions, requires further research to achieve efficient and effective learning. Further research is crucial to develop mechanisms to assist agents in exploring the search space effectively with good stability performance, especially in multi-agent collaborative systems.

Enhancing the exploration efficiency of search-based algorithms for effective trajectory optimisation in time-critical applications is essential. This can be achieved by modifying the training process of data samples during the offline pre-training phase [18] and using tuning probabilities for interaction with the environment to balance exploitation and exploration [18, 19]. Future research is needed to develop algorithms with fast convergence that can achieve optimal or near-optimal solutions with enhanced exploration capabilities.

4.1.7 Stability

Stability in RL refers to how consistently an algorithm performs when subjected to minor perturbations, as well as its ability to be documented and reproduced reliably. In online RL, stability is particularly important as it should operate with minimal monitoring, adapting in real-time to changing environments. Furthermore, RL algorithms need to generalise well across various real-world scenarios and perform effectively in both simulated test beds and real-world applications [20].

High stability is essential for DRL algorithms. Algorithms like DQN, while effective, can suffer from instability and poor convergence. Ensuring stability requires careful engineering of the algorithms to prevent underestimation and overestimation [21]. Regular updates to the target network and managing parameter changes during learning are crucial [11] for maintaining a clear trend in the learning process. Prioritizing robust algorithms and ensuring stable policy learning with low latency are key to achieving reliable results.

4.1.8 Choice of Parameters and Hyperparameters

Hyperparameters are fixed values that must be set before the learning process begins. In contrast, other parameters, such as the weight coefficients in a deep neural network, are learnable and derived through training. Hyperparameters, such as learning rate and batch size, are preset before training and significantly impact the performance of machine learning models [22]. The correct setting of hyperparameters is crucial, as only a few parameters typically contribute to most of the performance variation [23]. In UAV-assisted IoT applications, the learning rate is one of the most influential hyperparameters, and careful tuning is essential. Other important hyperparameters include exploration probability, loss coefficient, discount factor, number of neurons, and number of hidden layers [2]. Some of these parameters are directly related to the architecture and optimisation of the neural network.

Optimising neural network parameters with suitable loss functions is key to achieving optimal Q-function performance in DRL [8]. However, DRL involves numerous parameters that require extensive data and computational resources for training, making real-time deployment challenging, especially in AoI-sensitive applications with limited resources. While DRL often outperforms heuristic algorithms [10], it requires careful tuning of both parameters and hyperparameters to ensure optimal performance. For instance, the learning rate, which governs the speed and quality of learning, must be carefully managed to prevent slow convergence or instability [37].

4.2 AoI-Related Challenges in UAV-IoT Architecture

This section discusses some of the popular AoI-related challenges for the UAV-IoT architecture. Different from the previous section, the issues discussed are not peculiar to DRL-based solutions.

4.2.1 Reliability of Sensor Nodes' Connections

The reliability of communication between SNs and their respective cluster heads is often assumed but not guaranteed, especially because SNs are typically battery-constrained. The same challenge applies to communication between SNs and UAVs, where reliability varies depending on the specific scenario studied or deployed. Effective placement of SNs can improve connection reliability, as well as the strategic placement of UAVs. Moreover, it is crucial to prioritise protocol design and optimise IoT energy consumption to ensure timely and reliable information delivery [25]. This is because energy efficiency directly impacts communication reliability, making it essential to monitor and manage SN energy levels during data collection rounds. If SNs run out of energy during transmission, the data may not reach its intended destination. Therefore, considering end-to-end reliability is vital in UAV-assisted IoT communication. Efficient communication for SNs could require channel-aware routing strategies to improve communication reliability, both in terrestrial and underwater IoT networks. In fact, the reliability problem in underwater IoT is even more pronounced.

4.2.2 Optimal Scheduling Policies

Developing optimal scheduling policies for UAVs and terrestrial sensors is critical to maintaining minimal information age [25]. However, this is challenging when UAVs have no control over SN scheduling [26] or update packet arrivals. Queue-based policies are common [18] but may face difficulties when applied to UAV-aided IoT data collection architectures. Particularly, the access of SNs and the availability of data in their buffers could be highly unpredictable. Therefore, optimal queuing policies that prioritise crucial packets need to be developed. Optimal scheduling should be considered alongside several target objectives in UAV-assisted IoT networks, depending on the scenario being studied. These objectives could include power allocation, optimal user association, resource allocation, fairness optimisation, and UAV placement optimisation. In addition, these should also take into consideration the target application, which could vary from agriculture, disaster and pandemic management, environmental monitoring, industrial monitoring, and emergency rescue.

4.2.3 Traffic Generation Pattern Adaptation

IoTDs generate varying traffic patterns based on their sensing applications. Traffic patterns may vary significantly from uniform to bursty depending on the application scenario considered. Three common traffic patterns in IoT communication are Periodic Update (PU), Event-Driven (ED), and Payload Exchange (PE). Many IoT applications involve a combination of these patterns. For instance, a river monitoring system sends periodic measurements of water levels to a central server (PU). If the water level exceeds a certain threshold, an alarm may be triggered (ED). In response to this event, pictures might be sent to the server, or a data stream could be initiated to provide real-time updates. PU involves the periodic sending of update reports to a central entity at regular intervals, such as every second, with small packets of constant size. An example of this is remote health monitoring or smart meter reading. ED traffic, on the other hand, is sporadic and occurs only when a specific event is detected, such as a sensing device triggering an alert when a threshold is exceeded. The data size may vary depending on the application and the amount of information related to each event. This type of traffic is often real-time, especially when immediate action is needed, such as a security alarm in a surveillance system or health emergencies and disaster alerts. Finally, Payload Exchange involves the transfer of larger amounts of data, typically after an event triggers the need for more detailed information. For example, after a security alarm is activated, a video stream from a surveillance camera may be initiated to assess the situation in real time [27, 28]. In UAV positioning and trajectory planning, it is important to consider these traffic patterns alongside other optimisation objectives. This would ensure that the overall network performance is well optimised regardless of the traffic patterns. Offline trajectory planning is common but not suitable for dynamic environments where sudden changes are common. Online trajectory planning, which accounts for unknown activation patterns of IoTDs, for example, poses a significant challenge and can be further explored [15].

4.2.4 Choice of Algorithms and Trade-Offs

The selection of algorithms for UAV operations is crucial for achieving stability, convergence, and efficiency. Different algorithms have unique trade-offs, and choosing the most suitable one is not trivial. For instance, Deep Q-Network (DQN) algorithms, while promising, suffer from issues like poor convergence and instability. Algorithms like Proximal Policy Optimisation (PPO) offer fast convergence [29]. However, it is quite sensitive to changes in initial strategies [30]. The choice of algorithm impacts training time, energy consumption, computational requirements, and the reliability of solutions. Thus, it is important to select and deploy algorithms based on specific performance measures and application requirements.

Various ML techniques address AoI minimisation in UAV-assisted IoT, each with strengths and limitations. While DRL is popular, it faces issues like update instability and sensitivity to step size parameters [31]. RL, although effective for UAV trajectory design due to its autonomous learning [32], it suffers from long convergence times due to its trial-and-error mechanism, increasing computational complexity [32]. Managing multi-UAV scenarios where parts of the environment are unobservable further complicates the learning process. As opposed to learning-based schemes that do not deploy artificial neural networks, deep learning can aid in overcoming state-action space challenges, providing a level of generalizability [33].

4.2.5 Nature of the Optimisation Problem

The nature and scale of the network significantly influence the choice of optimisation algorithms. Large network architectures often challenge meta-heuristics, particularly in maintaining quality solutions. AoI minimization problems, which are frequently transformed into Markov Decision Process (MDP) problems, generally require dynamic programming [34] or linear programming solutions—methods that demand prior environmental knowledge. The curse of dimensionality in large-scale problems calls for computationally efficient algorithms, especially when dealing with diverse and numerous optimisation parameters. Moreover, not all problems lend themselves well to MDPs; while partially observable MDPs might be more suitable in some cases, other problems are better addressed using traditional heuristic or meta-heuristic algorithms such as ant colony optimisation, ant colony optimisation and genetic algorithms. Additionally, for certain problems, a model-based RL solution might be more appropriate, as it can be challenging to abstract them into the model-free RL framework. As such, the choice of algorithm depends largely on several factors such as the nature of the optimisation problem, the scale of the network, and the dimension of the problem.

4.2.6 Optimisation Algorithms

Intractable objective functions complicate problem-solving [31], often necessitating the formulation of tractable approximations or the relaxation of constraints. Combining traditional optimisation techniques with machine learning approaches can enhance UAV trajectory planning. However, ensuring timely algorithm convergence and avoiding local minima may require improvements to metaheuristic methods. While neural networks show promise, their convergence remains challenging and often relies heavily on extensive simulations [31]. Future research should aim to further harness the capabilities of meta-reinforcement learning, federated learning, and multi-agent reinforcement learning frameworks to optimise UAV-assisted IoT networks.

4.2.7 Environmental Challenges

Noise and interference, caused by the presence of other drones and environmental factors, present significant challenges. Although often overlooked, these issues impact the accuracy of experimental results, as simulated conditions differ markedly from real-world environments. For instance, UAVs often lack precise knowledge of their location, altitude, obstacle density, and channel conditions [35]. As a result, the randomness associated with Line-of-Sight (LOS) and Non-Line-of-Sight (NLOS) conditions must be carefully considered during system design [35]. Additionally, accounting for imperfect channel state information helps align simulation results more closely with real-world observations. To further mitigate real-environment limitations, deploying multiple antennas can enhance reliability. Equipping UAVs and SNs with multiple antennas, particularly in RIS-assisted UAV networks- could significantly improve network performance, despite introducing additional complexity [15, 36].

4.2.8 Proper Choice of Channel Models

The validity of the findings in simulations for investigating UAV deployment hinges on the accuracy of the channel models used. Real-world conditions, such as obstructions and varying channel quality, differ significantly from idealised models like the free-space channel model [31]. Accurate channel models, derived from real-world implementations, are essential for optimising UAV trajectories, stopping points, and transmission times. Inaccurate models can lead to performance degradation due to blockages [31]. Researchers often assume successful packet delivery [25], yet real-world conditions can adversely affect data quality. Therefore, precise channel modeling is crucial to predict the performance and thus minimise the Age of Information in UAV-assisted IoT applications.

4.2.9 Security Threats

Security is a critical concern in UAV operations, as UAV communication is inherently vulnerable to a range of attacks and security breaches. Insufficient authentication or lack of authentication between UAVs, SNs, and IoT devices can create significant vulnerabilities, enabling malicious activities and data compromise. Despite the growing importance of security in UAV-assisted IoT networks, few studies have comprehensively examined its impact on specific performance metrics, such as Age of Information. For example, a Sybil attack could mislead a UAV into collecting data from compromised IoT devices, which would increase AoI and degrade system performance. While implementing robust security measures (such as encryption) is essential, it presents challenges due to the limited processing power of UAVs and IoT/WSN nodes. Consequently, addressing these security threats remains a complex yet necessary task. Moreover, leveraging techniques like friendly jammers could enhance security by ensuring the confidentiality of UAV-IoT communication and preventing eavesdropping on sensitive data exchanges.

4.2.10 Fine-Grained UAV Assumptions

The study of UAV-assisted IoT architecture, whether through analytical methods or simulations, and the planning of UAV deployments in real applications require several key assumptions. These assumptions may include factors such as mobility, frequency of trips, acceleration or constant velocity, altitude, energy consumption, network coverage, payload size, and the type of UAV being used [1]. Additionally, many practical assumptions related to UAVs in the context of AoI minimization in WSN/IoT still warrant further exploration. For example, the impact of controlling UAVs through a Base Station (BS), the optimisation of drone fleet sizes based on task-specific requirements, and the deployment of heterogeneous UAVs with varying capabilities all present opportunities for deeper study. Furthermore, the effects of UAV capacity constraints, such as maximum payload limits and flight endurance, need to be investigated in greater detail. Real-world deployments often involve more complex scenarios than those typically considered in academic studies, such as dynamic SN distributions, multi-center data management, and environmental factors like wind. Addressing these complexities through further research will help improve UAV deployment strategies for IoT applications.

4.2.11 Network Architecture and Practical Constraints

Network architecture issues, particularly hybrid architectures where UAVs collect data from both SNs/IoT devices and collection points (CPs), require further optimisation. In some scenarios, UAVs must also assign tasks to IoT devices, making optimal task assignment strategies critical. Collaborative deployments involving multiple UAVs, including UAV swarms, demand careful interaction optimisation to guarantee timely data collection. Furthermore, fine-grained considerations for specific use cases, such as special missions in remote areas, are important, even if rare. Moreover, the impact of no-fly zones, especially in large-scale, continuous deployments, poses significant operational challenges that merit further study. Architectures requiring charging methods such as wireless power transfer, simultaneous wireless and information transfer, laser charging, energy harvesting from the environment, power line charging for drones, and tethered UAVs are potential solutions for real-life scenarios. Investigating these methods can improve the prediction and modeling of UAV-assisted IoT performance. Additionally, the practical applications of multi-tiered network architectures, the effects of UAV acceleration and deceleration, and the mismatch between theoretical estimates and actual field results warrant more in-depth investigation.

4.2.12 Real Deployments

Investigations targeted at application-specific real-world UAV deployment are highly required in this domain [38]. Practical implementation of DRL in UAV operations, especially for AoI-sensitive applications, requires realistic assumptions and close-to-practice system models. Real-world scenarios involve complex challenges like precise navigation to avoid collisions, environmental disturbances, and several uncertainties. The gap between simulation and real deployments poses significant challenges [39], and the successful transfer of DRL research from simulation to real environments needs thorough examination. The impact of practical environmental conditions like wind, rain, and dust must be studied [38] and put into consideration due to their practical implications on the system. Similarly, factors like hardware robustness, fault tolerance, and sample inefficiency must be addressed to ensure reliable deployment.

4.3 Summary

This chapter covers some of the critical challenges in implementing DRL for AoI-aware UAV-IoT applications. These include convergence, complexity, overhead, state-action space dimension, exploration, and stability concerns. It also discusses some important challenges in minimising the AoI in the UAV-IoT architectures, which covers issues involved in the simulations of the design models with the real deployment issues. Reliability of network connection, traffic generation pattern adaptation, security threats, UAV assumptions, environmental challenges, and choice of parameters, as well as hyperparameters, are some of the major concerns that need to be addressed while performing simulations of AoI optimisation in the UAV-assisted WSN/IoT networks. Eventually, the discussion in this chapter provides researchers with avenues for further research work in this field.

References

1. O.A. Amodu, U.A. Bukar, R.A.R. Mahmood, C. Jarray, M. Othman, Age of information minimization in UAV-aided data collection for WSN and IoT applications: a systematic review. J. Netw. Comput. Appl. 103652 (2023)
2. O.A. Amodu, C. Jarray, R.A. Raja Mahmood, H. Althumali, U. Ali Bukar, R. Nordin, N.F. Abdullah, N. Cong Luong, Deep reinforcement learning for AoI minimization in UAV-aided data collection for WSN and IoT applications: a survey. IEEE Access **12**, 108000–108040 (2024)
3. S.B. Prathiba, G. Raja, K. Dev, N. Kumar, M. Guizani, A hybrid deep reinforcement learning for autonomous vehicles smart-platooning. IEEE Trans. Veh. Technol. **70**(12), 13340–13350 (2021)
4. B. Feng, Z. Liu, G. Huang, C. Guo, Robust federated deep reinforcement learning for optimal control in multiple virtual power plants with electric vehicles. Appl. Energy **349**, 121615 (2023)
5. I. Jang, H. Kim, D. Lee, Y.-S. Son, S. Kim, Knowledge transfer for on-device deep reinforcement learning in resource constrained edge computing systems. IEEE Access **8**, 146588–146597 (2020)
6. Y. Yang, L. Juntao, P. Lingling, Multi-robot path planning based on a deep reinforcement learning DQN algorithm. CAAI Trans. Intell. Technol. **5**(3), 177–183 (2020)
7. M. Camelo, M. Claeys, S. Latré, Parallel reinforcement learning with minimal communication overhead for IoT environments. IEEE Internet Things J. **7**(2), 1387–1400 (2019)
8. X. Li, J. Li, D. Liu, Energy-efficient UAV trajectory design with information freshness constraint via deep reinforcement learning. Mob. Inf. Syst. **2021**, 1–9 (2021)
9. X. Xiang, S. Foo, Recent advances in deep reinforcement learning applications for solving partially observable markov decision processes (pomdp) problems: part 1-fundamentals and applications in games, robotics and natural language processing. Mach. Learn. Knowl. Extr. **3**(3), 554–581 (2021)
10. R.N. Boute, J. Gijsbrechts, W. Van Jaarsveld, N. Vanvuchelen, Deep reinforcement learning for inventory control: a roadmap. Eur. J. Oper. Res. **298**(2), 401–412 (2022)
11. N.P. Farazi, B. Zou, T. Ahamed, L. Barua, Deep reinforcement learning in transportation research: a review. Transp. Res. Interdiscip. Perspect. **11**, 100425 (2021)
12. A.Y. Majid, S. Saaybi, V. Francois-Lavet, R.V. Prasad, C. Verhoeven, Deep reinforcement learning versus evolution strategies: a comparative survey. IEEE Trans. Neural Netw. Learn. Syst
13. E. Marchesini, D. Corsi, A. Farinelli, Genetic soft updates for policy evolution in deep reinforcement learning, in *International Conference on Learning Representations*, (2020)
14. A. Al-Hilo, M. Samir, M. Elhattab, C. Assi, S. Sharafeddine, RIS-assisted UAV for timely data collection in IoT networks. IEEE Syst. J. **17**(1), 431–442 (2023). https://doi.org/10.1109/JSYST.2022.3215279

15. M. Samir, M. Elhattab, C. Assi, S. Sharafeddine, A. Ghrayeb, Optimizing age of information through aerial reconfigurable intelligent surfaces: a deep reinforcement learning approach. IEEE Trans. Veh. Technol. **70**(4), 3978–3983 (2021)
16. J. Hao, T. Yang, H. Tang, C. Bai, J. Liu, Z. Meng, P. Liu, Z. Wang, Exploration in deep reinforcement learning: from single-agent to multiagent domain. IEEE Trans. Neural Netw. Learn. Syst
17. P. Cai, H. Wang, Y. Sun, M. Liu, DQ-GAT: towards safe and efficient autonomous driving with deep Q-learning and graph attention networks. IEEE Trans. Intell. Transp. Syst. **23**(11), 21102–21112 (2022)
18. C. Zhou, H. He, P. Yang, F. Lyu, W. Wu, N. Cheng, X. Shen, Deep RL-based trajectory planning for AoI minimization in UAV-assisted IoT, in *2019 11th International Conference on Wireless Communications and Signal Processing (WCSP)*, (IEEE, 2019), pp. 1–6
19. M.A. Abd-Elmagid, A. Ferdowsi, H.S. Dhillon, W. Saad, Deep reinforcement learning for minimizing age-of-information in UAV-assisted networks, in *2019 IEEE Global Communications Conference (GLOBECOM)*, (IEEE, 2019), pp. 1–6
20. A.L. Trella, K.W. Zhang, I. Nahum-Shani, V. Shetty, F. Doshi-Velez, S.A. Murphy, Designing reinforcement learning algorithms for digital interventions: pre-implementation guidelines. Algorithms **15**(8), 255 (2022)
21. V. Mai, Reinforcement learning applied to the real world: uncertainty, sample efficiency, and multi-agent coordination
22. X. Dong, J. Shen, W. Wang, L. Shao, H. Ling, F. Porikli, Dynamical hyperparameter optimization via deep reinforcement learning in tracking. IEEE Trans. Pattern Anal. Mach. Intell. **43**(5), 1515–1529 (2019)
23. F. Hutter, H.H. Hoos, K. Leyton-Brown, An efficient approach for assessing hyperparameter importance, in *International Conference on Machine Learning*, (PMLR, 2014), pp. 754–762
24. O.A. Amodu, R. Nordin, C. Jarray, U.A. Bukar, R.A. Raja Mahmood, M. Othman, A survey on the design aspects and opportunities in age-aware UAV-aided data collection for sensor networks and internet of things applications. Drones **7**(4), 260 (2023)
25. M.A. Abd-Elmagid, H.S. Dhillon, Average peak age-of-information minimization in UAV-assisted IoT networks. IEEE Trans. Veh. Technol. **68**(2), 2003–2008 (2018)
26. A. Ferdowsi, M.A. Abd-Elmagid, W. Saad, H.S. Dhillon, Neural combinatorial deep reinforcement learning for age-optimal joint trajectory and scheduling design in UAV-assisted networks. IEEE J. Sel. Areas Commun. **39**(5), 1250–1265 (2021)
27. N. Nikaein, M. Laner, K. Zhou, P. Svoboda, D. Drajic, M. Popovic, S. Krco, Simple traffic modeling framework for machine type communication, in *ISWCS 2013; The Tenth International Symposium on Wireless Communication Systems*, (VDE, 2013), pp. 1–5
28. T. Garrett, S. Dustdar, L.C.E. Bona, E.P. Duarte, Traffic differentiation on internet of things, in *2018 IEEE Symposium on Service-Oriented System Engineering (SOSE)*, (IEEE, 2018), pp. 142–151
29. M. Sherman, S. Shao, X. Sun, J. Zheng, Optimizing AoI in UAV-RIS-assisted IoT networks: off policy versus on policy. IEEE Internet Things J. **10**(14), 12401–12415 (2023)
30. X.Y. Xu, Y.Y. Chen, T.R. Liu, TD3-BC-PPO: twin delayed DDPG-based and behavior cloning-enhanced proximal policy optimization for dynamic optimization affine formation. J. Frankl. Inst. **361**(12), 107018 (2024)
31. M. Samir, C. Assi, S. Sharafeddine, A. Ghrayeb, Online altitude control and scheduling policy for minimizing AoI in UAV-assisted IoT wireless networks. IEEE Trans. Mob. Comput
32. Z. Li, P. Tong, J. Liu, X. Wang, L. Xie, H. Dai, Learning-based data gathering for information freshness in UAV-assisted IoT networks. IEEE Internet Things J
33. Z. Qin, X. Zhang, X. Zhang, B. Lu, Z. Liu, L. Guo, The UAV trajectory optimization for data collection from time-constrained IoT devices: a hierarchical deep q-network approach. Appl. Sci. **12**(5), 2546 (2022)
34. M. Yi, X. Wang, J. Liu, Y. Zhang, B. Bai, Deep reinforcement learning for fresh data collection in UAV-assisted IoT networks, in *IEEE INFOCOM 2020-IEEE Conference on Computer Communications Workshops (INFOCOM WKSHPS)*, (IEEE, 2020), pp. 716–721

35. M. Sun, X. Xu, X. Qin, P. Zhang, Aoi-energy-aware UAV-assisted data collection for IoT networks: a deep reinforcement learning method. IEEE Internet Things J
36. D.-H. Tran, V.-D. Nguyen, S. Chatzinotas, T.X. Vu, B. Ottersten, UAV relay-assisted emergency communications in IoT networks: resource allocation and trajectory optimization. IEEE Trans. Wirel. Commun. **21**(3), 1621–1637 (2021)
37. M. Yi, X. Wang, J. Liu, Y. Zhang, R. Hou, Deep reinforcement learning for energy-efficient fresh data collection in rechargeable UAV-assisted IoT networks, in *2023 IEEE Wireless Communications and Networking Conference (WCNC)*, (IEEE, 2023), pp. 1–6
38. A.T. Azar, A. Koubaa, N. Ali Mohamed, H.A. Ibrahim, Z.F. Ibrahim, M. Kazim, A. Ammar, B. Benjdira, A.M. Khamis, I.A. Hameed et al., Drone deep reinforcement learning: a review. Electronics **10**(9), 999 (2021)
39. S. Zhang, Y. Li, Q. Dong, Autonomous navigation of UAV in multi-obstacle environments based on a deep reinforcement learning approach. Appl. Soft Comput. **115**, 108194 (2022)

MIX
Papier aus verantwortungsvollen Quellen
Paper from responsible sources
FSC® C105338
FSC
www.fsc.org

If you have any concerns about our products,
you can contact us on
ProductSafety@springernature.com

In case Publisher is established outside the EU,
the EU authorized representative is:
Springer Nature Customer Service Center GmbH
Europaplatz 3, 69115 Heidelberg, Germany

Printed by Libri Plureos GmbH
in Hamburg, Germany